Rudolf Kruse
Jörg Gebhardt
Rainer Palm (Eds.)

Fuzzy Systems in Computer Science

Artificial Intelligence
Künstliche Intelligenz

edited by Wolfgang Bibel and Walther von Hahn

Artificial Intelligence aims for an understanding and the technical realization of intelligent behaviour. The books of this series are meant to cover topics from the areas of knowledge processing, knowledge representation, expert systems, communication of knowledge (language, images, speach, etc.), AI machinery as well as languages, models of biological systems, and cognitive modelling.

In English:

Automated Theorem Proving
by Wolfgang Bibel

Parallelism in Logic
by Franz Kurfeß

Relative Complexities of First Order Calculi
by Elmar Eder

Fuzzy Sets and Fuzzy Logic
Foundations of Application – from a Mathematical Point of View
by Siegfried Gottwald

Fuzzy Systems in Computer Science
by Rudolf Kruse, Jörg Gebhardt and Rainer Palm

Selected titles in German:

Prolog
by Ralf Cordes, Rudolf Kruse, Horst Langendörfer, Heinrich Rust

Wissensbasierte Systeme
by Doris Altenkrüger and Wilfried Büttner

Logische Grundlagen der Künstlichen Intelligenz
by Michael R. Genesereth and Nils J. Nilsson

Logische und Funktionale Programmierung
by Ulrich Furbach

Wissensrepräsentation und Inferenz
by Wolfgang Bibel (with St. Hölldobler and T. Schaub)

Deduktive Datenbanken
by Armin B. Cremers, Ulrike Griefahn and Ralf Hinze

Neuronale Netze und Fuzzy-Systeme
by Detlef Nauck, Frank Klawonn and Rudolf Kruse

Rudolf Kruse
Jörg Gebhardt
Rainer Palm (Eds.)

Fuzzy-Systems in Computer Science

Verlag Vieweg, P.O. Box 58 29, D-65048 Wiesbaden

Vieweg is a subsidiary company of Bertelsmann Professional Information.

Softcover reprint of the hardcover 1st edition 1994

Printed on acid-free paper

ISBN-13: 978-3-322-86826-8 e-ISBN-13: 978-3-322-86825-1
DOI: 10.1007/978-3-322-86825-1

Preface

In recent years it has become apparent that fuzzy systems provide useful tools for obtaining greater generality, higher expressive power, and more convenient concepts for modelling imprecision and uncertainty in many real world applications in order to achieve tractability and low cost realizations without affecting the overall quality of products. For this reason fuzzy systems have an increasing impact in the realms of artificial intelligence, information processing, diagnostics, intelligent control, optimization techniques, decision analysis, and related fields. New and active areas of research emerged, equipped with many practical applications and interesting theoretical problems that have not been solved yet. In this connection several German research projects and working groups in Computer science have been formed that focus on various topics of fuzzy systems. Considering the need for a regular forum where the corresponding work can be discussed by specialists, the research group "Fuzzy Systems" was founded in October 1993 as a part of the German Society of Computer Science (GI).
The first important activity of this group referred to the organization of the workshop "Fuzzy Systems - Management of Uncertain Information" in Braunschweig (October 20-22, 1993) with about 120 attending participants. Invited speakers were Abe Mamdani, the founder of fuzzy control, and Didier Dubois, the designed World President of the International Fuzzy Systems Association (IFSA). All major German research groups on fuzzy systems contributed to this workshop.

In this book we want to address some essential topics that were discussed at the mentioned workshop. The whole presentation is organized as follows:
The first paper gives an overview about the historical development of fuzzy systems in Germany. Then the book is partitioned into the five following chapters:

- Fuzzy Control
- Fuzzy Neuro Systems
- Fuzzy Systems in AI
- Theory of Fuzzy Systems
- Fuzzy Classification

These chapters reflect basic trends and recent results in fuzzy systems methodology. The organization of each section is uniform: It starts with an authoritative introduction to the main issues of the respective field of research and applications, involving the actual state of the art. The preceeding papers in a section address recent deliverables and present new ideas regarding the improvement of fuzzy systems.

Since theory and application of fuzzy systems is highly interactive in different fields, the material addresses practitioners and scientists in computer science as well as control engineering, the natural sciences, and mathematics.

The editors wish to thank the authors who contributed their work to this book. We also express our gratitude to Reinald Klockenbusch from Vieweg Verlag for his support, and to our students Heiner Bunjes and Roland Stellmach for their excellent assistance in putting together the final manuscript.

Rudolf Kruse
Jörg Gebhardt
Rainer Palm

June, 1994.

Contents

1

Fuzzy Systems in Germany: Historical Remarks

1. Fuzzy Systems in Germany: Historical Remarks

Hans–Jürgen Zimmermann

Fuzzy Set Theory was recognized by some scientists in Germany already at the beginning of the seventies. While control engineers in Great Britain developed the concept of a fuzzy controller and showed that it worked, on the Continent, research was more done in the mathematical areas and in particular in Operations Research where the first institutional working group (European working group of fuzzy sets) was established in 1976. In German universities research was predominantly in mathematical areas. At the Universities of Wuppertal and Mainz, for instance, research was performed in the areas of fuzzy topology and algebra while at the University of Braunschweig in the early eighties research on fuzzy measures and the interface between classical statistics and fuzzy set theory was done. At the Institute of Technology in Aachen (Aix-la-Chapelle) empirical and axiomatic research went on at that time. In the Chair for Operations Research, empirical as well as axiomatic basic research concerning operators and membership functions was started in 1972. This research overlapped with the development of "fuzzy linear programming" and its applications to multi–criteria–analysis and various other areas.

Even though some applications, in particular, of fuzzy linear programming can be found during these years, most of the activities in the area of fuzzy set theory were more or less in the academic field. Even here there were hardly any courses offered at that time. Classes in fuzzy set theory and its applications were offered on a regular basis since 1982 only at the Aachen Institute of Technology. This was accompanied by quite a number of master and Ph.D. theses between the middle of the seventies and the middle of the eighties. In the middle of the eighties research work shifted from basic research and fuzzification of Operations Research Methods to the application of fuzzy sets to knowledge based systems. Focus at that time was still the area of fuzzy control. The shell that was developed at the Institute and which was available around 1989, was called "FIT" (Fuzzy and Intelligent Techniques).

The scene changed drastically at the end of 1990, when one of the leading German Journals (highTech) published in two consecutive issues cover stories on "fuzzy logic". At the same time the German television broadcasted in the "Computer Club" existing fuzzy products from Japan (video camera, pattern recognition devises, etc.). In these publications and broadcasts, Aachen was mentioned as a place where people had been working in this area for quite a while. It is, therefore, not surprising that within the first two or three months of 1991 we received approx. 150 requests for cooperation and help. Most of the requests described the problems for which they wanted to design a fuzzy solution quite well and about 70 % of those problems were really well suited for the applications of fuzzy technology.

In May of 1991, the Journal that had published the first articles, started a road show, i.e. a series of introductionary seminars aimed at high management levels. These seminars took place in Munich, Stuttgart, Frankfurt, Düsseldorf and Aachen (within one week) and each of the seminars had more than 60 participants. At the CeBit and the Industrial Fair in Hannover, the first exhibits of fuzzy technology attracted an unexpected number of visitors which just wanted to know what that was, "fuzzy logic". Exhibited was a fuzzy car, i.e. a small model car with a top speed of 80 km/h that navigated without any external communication, just on the basis of a built in fuzzy knowledge based system, built at the Institute of Technology in Aachen. Also some presentations by Omron, a Japanese firm that tried to push fuzzy technology in Germany first. Still in 1991, the first fuzzy tool (Fuzzy Tech) was offered by INFORM in Aachen. This had been possible because "FIT", as mentioned above, had already been completed and could quickly be changed into a commercial case tool which now ranges under the three first world-wide sales-wise.

The increasing interest in the public sparked off interest in fuzzy technology in numerous German Universities and Institutes of Technology as well as in small and large German companies. Siemens, for instance, started a "Task Force fuzzy technology" and developed in cooperation with an American software-house and with German companies fuzzy software, fuzzy hardware and applications. In particular, numerous Chairs for Control Engineering became interested in the Japanese applications for fuzzy control which seemed to be quite attractive.

Four German professional societies (Computer Science, Operations Research, Electrical Engineering, Mechanical Engineering) started working groups; one of the societies even two, a theoretical and an applied one. These working groups already met several times in 1991, partly by organizing their group meetings as symposia on fuzzy technology.

The governmental acceptance of this technology and the public financial support failed, however, to develop. Most of the big German companies started groups that began to develop fuzzy products, fuzzy tools or fuzzy methods, but it seemed to be very difficult for these companies to cooperate on a pre-competitive level. An attempt to set up an European institution as counterpart to the LIFE Institute in Japan which was founded of 49 big Japanese companies, and to BISC (Berkeley Soft Computing Initiative) was started. This seemed to fail due to long decision making times in the enterprises and their missing willingness to any type of cooperation. It also became clear that the German magerial attitude differs significantly from the Japanese by being much more hesitant in investing in new technologies before the success had been proven by competitors. This difference had already led in the eighties to the lead of Japanese companies in the fuzzy area.

Two important and very positive exceptions, however, happened still in 1991: In December 1991, after a very short time of preparation, the Ministry of Economic Affairs of the State North Rhine Westphalia decided to start the "Fuzzy Initiative North Rhine Westphalia" and at the same time the Europe an foundation ELITE (European Laboratory for Intelligent Techniques Engineer ing) could be founded in Aachen and started operation in January 1992. The Fuzzy Initiative North Rhine Westphalia is a project that carries almost 10 million Deutschmarks of financial support and aims primarily at technology transfer in the area of fuzzy technology. It also tries to provide help to potential customers or developers of fuzzy products by supporting activities in a demonstration center (in Dortmund), a consultation center in Aachen, data banks for literature, patents and projects, by subsidizing seminars and the development of fuzzy technology. The data bank for literature, projects, events, and suppliers of fuzzy products has, in the meantime, been turned into a commercial product and service (CITE) which is available worldwide, and ELITE is the prime proposer of quite a number of European Research projects in the area of intelligent technologies and a host of several leading visiting scientists from around the world. It can be expected that Fuzzy technology will also play a major part in the Fourth Framework Programme of the European Community starting in 1995.

The development in 1992 followed that of 1991, but already in a considerably larger scope. There were at least 20–30 seminars and symposia which drew between 20 and 500 participants and which normally lasted between one and three days. On the three best-known industrial fairs CeBit, InterKama and the Industrial Fair in Hannover, quite a number of fuzzy products were already shown. The German "Fuzzy Initiative North Rhine Westphalia" also became effective. In the meantime it has established a scientific board, a technical board and, as the most important body, the "Fuzzy Club", which consists of approx. 500 members that are invited to different types of events regularly and that are the most effective means of a fast technology transfer from the rather scarce "fuzzy resources" to those who can use them. Actively involved in the "Fuzzy Initiative North Rhine Westphalia" are also all Chambers of Commerce and Industry of the State of North Rhine Westphalia. Of particular interest seem to be the "Anwendertreffen". These are symposia in which practitioners report about their new products and experiences for the sake of other practitioners. These meetings normally last two days; there are about 20 presentations and the usual audience consists of more than 100 participants. In 1992 there were four to five of these symposia in Germany and in 1993 there were at least six to seven of these events scheduled in different locations. It can, however, be observed that the number of participants in these events is decreasing. This is not necessarily an indication of decreasing interest in the fuzzy area, but to a certain extent an expression of consolidation. New products and developments cannot be generated that fast that several conferences per year can be filled with new and interesting applications and commercial interests may also supersede the desire of the speakers to present new developments in detail to the public. It should probably be mentioned that in 1993 a Japanese/North Rhine Westphalian symposium took place in Düsseldorf and a European/Japanese symposium in Berlin.

A new development started in September of 1993 with the first European Congress on Fuzzy and Intelligent Techniques (EUFIT '93). This congress drew approx. 500 participants of which 300 came from universities and the rest from industry and other research and administration institutions. When the Fuzzy–Neuro–Initiative of North Rhine Westphalia was officially opened in the framework of this conference the number of participants even grew to approx. 700. Roughly half of the participants came from Germany, about 10 % from abroad and the rest from other European countries. It included again – similar to other events of this type – an interesting exhibition of fuzzy products as well as of fuzzy tools. A growth of these exhibitions, in particular, indicate a growing acceptance of fuzzy technology in industry and in administration. Even though the majority of applications is still in the area of fuzzy control and other more technical areas, such as quality control, analytics, etc. applications in the management areas (production control, market segmentation, strategic planning, etc.) seem to appear in growing numbers.

This is particularly true for applications of fuzzy data analysis in very diverse areas. This development is certainly facilitated by a considerable number of very user friendly case tools (fuzzy control shells) which have been built in Germany during the past two years including the probably first tool for fuzzy data analysis. A certain maturity of this area is also documented by a surprisingly high number of books in German on fuzzy technology that came out during 1993/94: including the two or three translations from English into German more than 30 books have appeared in German until 1994.

In 1994 a certain synchronisation of the development in Europe, Japan and the USA occurs: conferences in this area in all three continents (including EUFIT '94) follow the same pattern by including, generally under the name "intelligent computing", the three areas of fuzzy technology, artificial neural nets, and evolutionary computing. This is certainly a very attractive development because these three areas have been cross fertilizing each other since the beginning of the 90ies and a closer cooperation of people and institutions in these three areas can certainly generate synergies and cut out double work.

2

Fuzzy Control

2.1. Fuzzy Control: An Overview

Hans Hellendoorn

Abstract

Fuzzy control is on its way to become an established control theory besides other modern control techniques. But there are several new trends, new directions in fuzzy control that should be observed. First of all, the trend from first generation to second generation systems. Secondly, application examples of fuzzy theories besides fuzzy control, such as fuzzy data analysis, fuzzy diagnosis, *etc.* Thirdly, the combination of fuzzy with other modern techniques like neural networks and genetic algorithms. Fourthly, the coming into existence of a design and development methodology for fuzzy control, similar to that of conventional control. Here also the combination of fuzzy control with other control techniques plays a role.

1 First and Second Generation Systems

"The fuzzy wave has reached Europe" someone wrote in the beginning of the nineties. In a relatively short period of time many applications with a label 'fuzzy' came to the market: household appliances, automobile components, automation systems, PLCs, traffic systems, etc. In particular Japanese market studies show an extremely wide variety of application areas. On an international workshop in Munich, 1992, E.H. Mamdani, who may be called the inventor of the fuzzy control theory, expressed the reason of this fuzzy wave as follows: "There is an abundance of open-loops in modern industry, in cameras, in cars, in video recorders, etc. These are all problems where man has to close to loop by hand, e.g., in focussing a camera. Japanese industry uses fuzzy control to close many of these open loops. Most often it concerns relatively simple problems, but there are so many small and relatively simple problems that actually demand to be solved." What happened in the late eighties and beginning nineties was exactly that. Fuzzy control was tested on many relatively simple problems. Compare, for example, the large number of publications on the inverted pendulum. The inverted pendulum problem is from a control point of view, or more striking, from an automation point of view, an extremely simple problem when compared with the control of a chemical process or a large electricity network. Therefore, in fuzzy control we have introduced the division in first and second generation fuzzy systems.

A first generation fuzzy system is characterized by the following properties.

1. Small-scale knowledge bases
Most first generation fuzzy control systems are two inputs one output systems, with usually three to seven fuzzy sets on each domain. This leads, for a two inputs system with five fuzzy sets on each domain, to potentially 25 rules, where usually only ten to fifteen rules are being used. The difference between such a system and systems as described by Dr. Meyer-Gramann in this volume are gigantic. There are no problems with automatic consistency or completeness proofs, the rule base can easily be represented by a simple look-up table on a PC-screen, and it can easily be checked which rules "fire" during run-time.

2. Purely fuzzy
Most first generation fuzzy systems are purely fuzzy, that means, they are stand alone systems and are not hybrid. They do not contain self learning, neural networks, chaos theory, or genetic algorithms components, neither are they integrated in a larger control or diagnosis environment. The inputs are usually clearly defined and crisp, i.e., there are no problems with stochastic uncertainties due to low quality sensors as described by Palm [Palm94].

3. Software-based
Usually, to develop a fuzzy controller or, more generally, a fuzzy system, software tools on PC are used. These development tools contain editors to define the domains, the fuzzy sets for the input and output domains, the rule bases, etc., and are able to translate such a fuzzy system into a programming language, usually C, but sometimes also to special microprocessor or PLC-languages. More elaborate fuzzy development tools like SIEFUZZY (Siemens), TILShell (Togai InfraLogic), FIDE (Aptronix), and FuzzyTech (Inform) also offer analysis and testing possibilities for the fuzzy system. The result of such a process is a purely software based fuzzy system. Most today fuzzy applications and prototypes are based on this principle. Problems can occur when the system has to be online adapted in real-time and the processor is not quick enough to allow this. In that case one may need special fuzzy hardware. Another problem may occur if, due to production costs, the hardware is very limited in

its performance, either in its memory, its speed, or both. In that case one has to count the costs where to make changes and to put away fuzzy components.

4. Min-max-centroid

In our book [Driankov/Hellendoorn/Reinfrank93] we have shown a great variety of fuzzy operations that can be used in fuzzy control. We showed there alternative inference methods (Mamdani and Gödel), several T-, S- and c-norms to implement *and*, *or*, and *not*, a variety of defuzzification operators, etc. An even greater collection of such operators can be found in [Dubois/Prade80] and many other publications from Dubois & Prade. Furthermore, many fuzzy systems, for example fuzzy data analysis systems or fuzzy diagnosis systems — both typical representatives of second generation fuzzy systems —, do use completely other operators and do not use defuzzification. Nevertheless, most of today fuzzy systems use only minimum and maximum operators for *and* and *or*, and use the center-of-sums [Hellendoorn/Thomas93] or centroid defuzzification method. Experience has shown that in larger fuzzy systems these simple mathematical operations do not satisfy.

5. Fuzzy control

Most problems that were solved in the past with the help of the fuzzy theory were control problems. Therefore, it is usual to talk about fuzzy control as a generic term, denoting that part of the fuzzy set theory that is used for applications. This means that there is no differentiation between, e.g., fuzzy control, fuzzy classification, fuzzy diagnosis, fuzzy expert systems, fuzzy data analysis, fuzzy image processing, etc. Second generation fuzzy systems tend in a direction away from pure control to more complex integrated systems (see Sect. 2)

6. Projecting

In conventional control theory there are many ways to build a control system. It has to be decided whether a PI or PD controller should be used, Ricatti-methods can be used, there are methods to solve delays and hysteresis, etc. But all methods have in common that in the end, after having implemented the system in a real time environment, an operator has tune the system. It would be an illusion that fuzzy controllers are so much better designed due to the rule-based approach that they do not need this last step. Nevertheless, in almost no earlier publication on fuzzy control application this problem is mentioned. Palm describes this problem extensively in his paper in this volume.

In this paper we want, as the title suggests, to deal with *new* trends in fuzzy control, with properties of second generation fuzzy systems. It will be clear now where extensions to the current theory can be expected. Fuzzy systems theory and applications can be considered as an amoeba that is growing in different directions. We will describe the following directions. In Section 2 we will focus on the application axis. We will consider there new kinds of applications and describe inherent problems that have to be solved. In Section 3 we will describe combinations of fuzzy systems with, e.g., neural networks or genetic algorithms and describe some possibilities in that area. In Section 4 we will describe design and development problems of more complex fuzzy systems. Furthermore, we will compare the design of fuzzy systems with that of conventional control and see how relatively young and unexperienced fuzzy is in comparison to her so much elder sister — but also more attractive.

2 New Application Trends in Fuzzy

When one considers lists of applications realized with the help of fuzzy then it is strange that everyone talks about *fuzzy control* although many projects have to do with classification and diagnosis. Fuzzy control is like the notion *fuzzy logic* being used in two ways. *Fuzzy logic* is on the one hand side being used as a generic term, denoting the whole fuzzy set theory from fuzzy control to fuzzy topology, on the other hand side, it is being used in its narrow sense denoting, say, approximate reasoning. In the same way *fuzzy control* is more and more being used as a generic term denoting that part of the fuzzy set theory that is used for applications as well as in its narrow sense, denoting closed loop fuzzy systems. Nevertheless, we believe it is better to differentiate between, e.g., fuzzy control, fuzzy classification, fuzzy diagnosis, fuzzy expert systems, fuzzy data analysis, fuzzy image processing, etc. In general, a fuzzy system can be described like in Fig. 1a. There is a process, e.g., a paper processing system or a car, a fuzzy system, and a development system, e.g., a PC-tool. The developer of the fuzzy system stands on the top of the figure. Furthermore, there may be operators, one is responsible for the input of the fuzzy system and the output of the process, the other one may deliver input to the process and obtains output from the fuzzy system. This needs not necessarily be a human operator, it can also be another fuzzy or non-fuzzy system delivering inputs or processing outputs. From this picture we can derivate several kinds of fuzzy systems, e.g., fuzzy

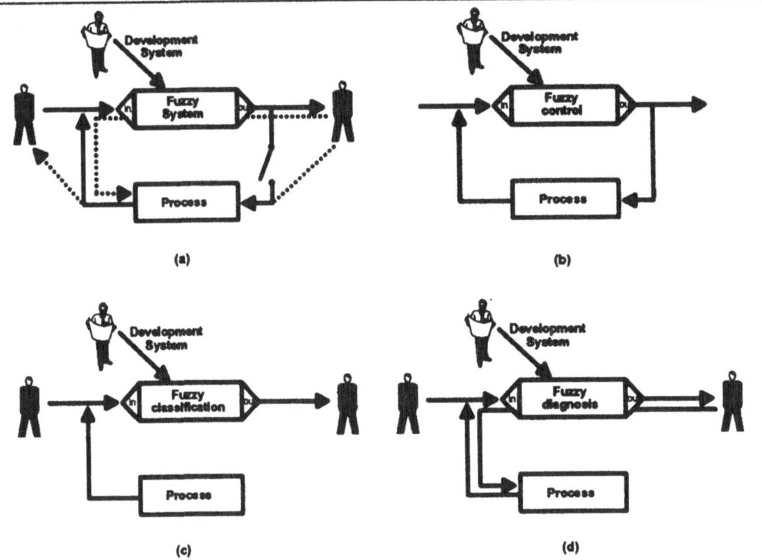

Figure 1: A generic fuzzy system.

controllers, fuzzy classificators and fuzzy diagnosing systems.

A *fuzzy control system* (Fig. 1b) is a closed loop system, so there is no operator available or the operator is part of the control loop. An example of a fuzzy control system within Siemens is the vacuum cleaner that adapts its motor power depending on the amount of dust on the floor and the kind of carpet. Another example is the torque optimizer in the anti-slip system that is used in trains and underground systems. Inputs are, among others, the velocity of the train and the resistance of the rails.

Classification is basically an ordering operation on metric and non-metric scales, i.e., based on expert opinions as well as on measured and usually prestructured data. Mathematically, classification is related to set theory rather than functions. In fuzzy classification, it is possible to deal with vagueness of the expert opinions and with badly defined model based relations, e.g., input-output relations. Classification systems are able to allocate unknown objects or system states to a number of classes. If one considers the different input vectors as axes in a high dimensional space frame, the theoretical basis of fuzzy classification can be considered as a pattern concept. So, a set of classes in a high dimensional input space is described by the classification model. This input space is described by properties of the objects and forms a kind of model. In fuzzy classification the properties may be described by linguistic values and the classes that exist in the high dimensional space frame need not be disjunct. The task of the system designer is to find out which properties or input signals are needed and play a significant role to classify the system. To a certain degree the classification process becomes better when more input signals are available (due to errors in the input signals and correlation too much input signals can cause the system to loose its generality) but in practical implementations the price of the product usually puts a natural limit on the number of input signals.

A *fuzzy classification system* (Fig. 1c) does not consist of a loop. It takes the input and output of some process and tells in what state the process is. This information can, of course, be used to control the process directly or to give the operator the opportunity to interfere. An example of a fuzzy classification system is the fuzzy washing machine, where some parameters of the washing machine are used to determine the amount and kind of laundry. The output of this classification is used to take a decision how to spin-dry or how to get optimal friction between the laundry. Another example is the fuzzy automatic transmission system. Using some sensors that are available in the car (e.g., from the ABS system, the power steering system, the motor control system, etc.) the classification system determines the state of the car (e.g., the car is loaded, the car is going uphill, etc.). So the pictures of the human beings on the left and right hand side of the fuzzy classification system in Fig. 1c will usually be other systems, delivering and processing the information. In a *fuzzy diagnosis system* (Fig. 1d) human beings play an explicit role. They may deliver input to the system when the system explicitly asks for additional information. Furthermore, they may ask the diagnosis system how it came to its conclusion or to give more details about the diagnosing procedure.

Fuzzy diagnosis systems are closely connected with fuzzy expert systems. The both differ from fuzzy classification systems due to the larger role of the human operator. Application areas of fuzzy diagnosis systems are usually large plants that can either only with great difficulties be described by exact algorithms or are difficult to model with conventional mathematical models. In the first case it often happens that the rule base becomes too large. Prof. Zadeh is true when he states that fuzzy expert systems use much less rules than conventional expert systems which improves the readability and helps to avoid inconsistency and incompleteness. It has to be stated that there exist many model

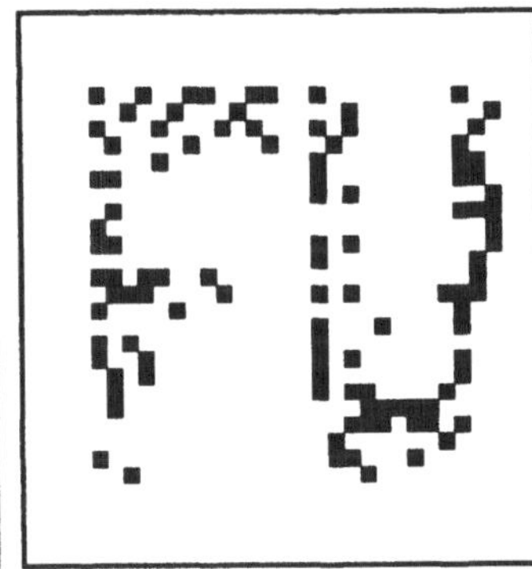

Figure 2: Two examples of clusters or patterns that can be recognized with human eyes. Left: sun, airplane, house, tree. Right: the characters F and U.

based diagnosis systems that perform very well. It is only in these cases where the problem is hard to model mathematically or otherwise that fuzzy should be used. An advantage of fuzzy diagnosis systems is that the knowledge of the operator can be used.

An important aspect of *diagnosis systems* is the presence of a user interface. Usually, the human operator can ask questions to the system and can order the system to explain why it got to a certain diagnosis. Another aspect is the way the results are presented to the user. Normally, defuzzification makes no sense in diagnosis systems. What is needed for these systems is a better theory of linguistic approximation to give the user understandable answers.

Fuzzy data analysis is more difficult than any of the aforementioned theories. The main goal of fuzzy data analysis or fuzzy cluster analysis [Bezdek81] is to search for structure in data. A subgoal is to reduce the complexity of the data structures. Fuzzy data analysis is like fuzzy diagnosis very similar to fuzzy classification. Nevertheless are there some differences which makes fuzzy data analysis a strict subclass of fuzzy classification. The best instrument for data analysis is the human eye. Clusters of data or patterns that can very easily be detected with the eyes may be extremely difficult to detect with automated methods (see Fig. 2). The first task of fuzzy data analysis therefore is to find features of the data that can help to cluster and identify the data. Here one can use fuzzy notions like 'close', 'compact', 'bulge', *etc.* After this each cluster has to be identified. For example, in character recognition, one has first to identify the individual characters, and, after that, one has to give names to each character. Each of these tasks has turned out to be extremely difficult. A problem with the fuzzy theory is to find the right fuzzy operators to define notions like 'similarity' or 'dissimilarity' of two fuzzy sets.

3 Fuzzy-Neural-Genetic-Chaos

Another trend in fuzzy control is the combination of fuzzy systems with other modern techniques like neural networks, genetic algorithms, and chaos theory. Several 'neural-fuzzy' products are already available on the market. Furthermore, the Korean company Goldstar lately presented a 'fuzzy-chaos-neural' washing machine. The difficulty with all these notions is that they can easily be used as buzz-words to make advertisement for products, but there is no guarantee that by combining two good techniques one obtains a new technique with the advantages from both. In the worst case, the new technique inherits exactly the disadvantages from its 'parents'.

In this section we will show some ways how to combine fuzzy systems and neural networks. Furthermore, we will show the use of genetic algorithms. Lastly, we will make some remarks about chaos theory and nonlinear dynamics.

There are many ways to combine neural networks (NN) and fuzzy systems (FS): FS preprocessing data for NN and vice versa, FS calculating parameters of NN and vice versa, FS delivering knowledge to prestructure NN, FS to interpret NN, NN to learn fuzzy rules and rule weights, *etc.* We will now discuss the four most used ways (Figure 3).

a. Use NN to generate input for FS (Fig. 3a)
There are many processes that are difficult to describe with mathematical models and where it makes sense to determine the strategy for the control with the help of a neural network and to use these values in fuzzy rules. The fuzzy rules then control the process or even deliver information to the actual control level of the plant where conventional controllers may be used. So the arrow in Fig. 3a) has a temporal meaning: first, the neural network comes into action, the the fuzzy system does its work. An example of this can be found in [Osaki et al.]. Another

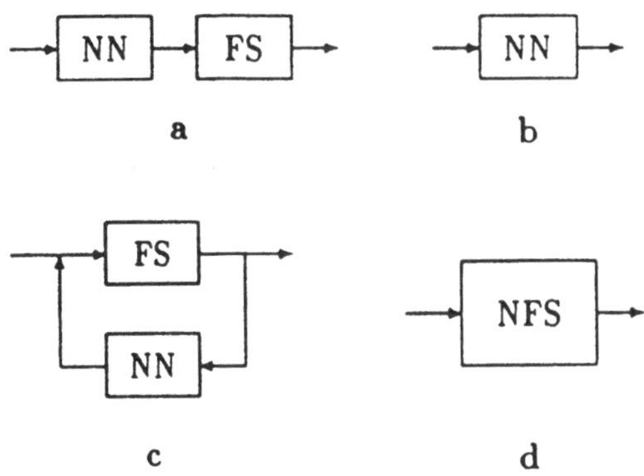

Figure 3: Four ways how neural networks and fuzzy systems can be combined.

example is the Siemens paper processing system that was installed in a plant in Caima (Portugal). Here, neural networks are used to calculate the cooking time and optimal pressure depending from the quality and kind of trees that come in. Afterwards, the several fuzzy systems control the process (cf. Fig. 4).

b. Use NN to generate parts of a FS (Fig. 3b)
Inputs to the neural network are numeric or linguistic data, the output are rules, parts of rules or membership functions. This approach benefits from the ability of neural networks to perform complex classification tasks. This method takes place once to build up the fuzzy system. The principle is illustrated in Fig. 5. A cluster in the input-output domain is recognized by a neural network. Then fuzzy sets on the input domain are either chosen out from a predefined set of fuzzy sets or determined by the neural network. Consecutively, rules are formed that can even have rule weights that depend on the quality of the classes, e.g. the compactness of a class. TILGen and the neural network tool in FuzzyTech are based on this principle. We have tested these systems with many practical examples in our laboratory. These systems work well in relatively simple two inputs one output systems. In more complex systems it turns out that other classification methods perform as well or even better.

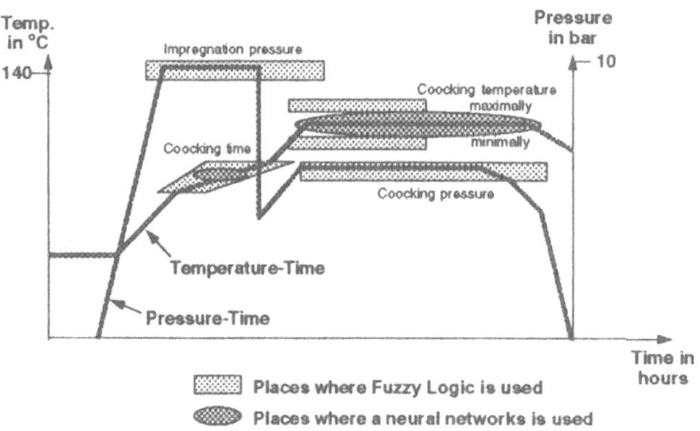

Figure 4: An overview of the role of fuzzy logic in the paper production process.

c. Use NN to adapt a given FS (Fig. 3c)
This means to introduce learning to a fuzzy system. Wang and Mendel [Wang/Mendel91] discuss an approach how network learning can directly be applied to fuzzy rules. Usually, in this case neural networks are used to learn an inverse model of the process by observing the input-output data. This inverse model is used to adapt parameters of the fuzzy system online. This means that the fuzzy system has the possibility to adapt itself during its life-time to the process on which it is operating, such taht it becomes optimally appropriate for its task. A disadvantage of this method is that the fuzzy system can learn wrong things. In the paper processing company such a system might learn during a long and dry summer that there exist only dry trees and will be pretty 'surprised' when after the first autumn showers wet trees are brought into the process! Here one needs new techniques to avoid drifting of the system.

d. Integrate FS and NN (Fig. 3d)
This approach is intriguing with respect to the combination of the advantages of the different approaches or the compensation of the disadvantages respectively. One possible way is to translate a fuzzy system

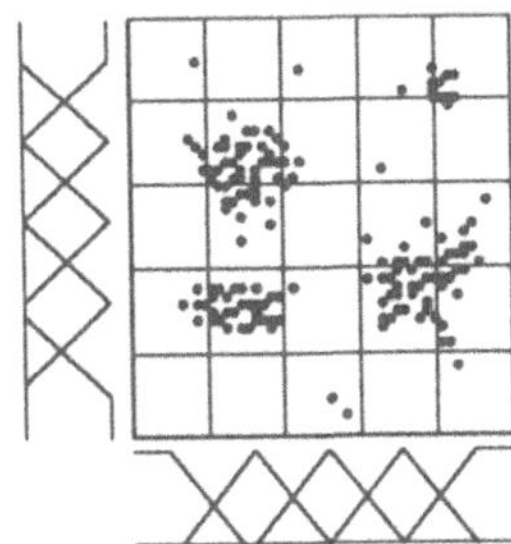

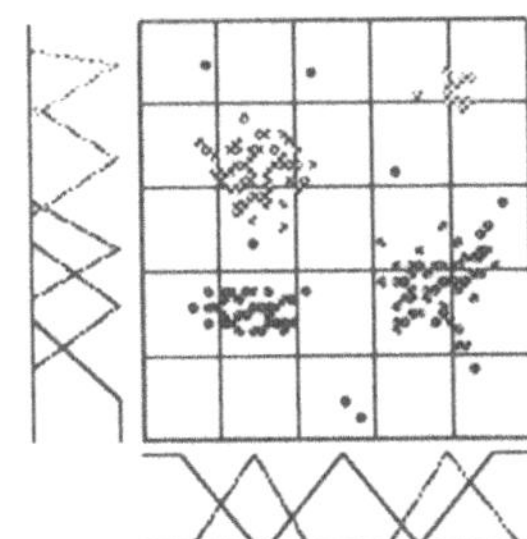

Figure 5: An example of the correspondence of the locations of clusters and those of the input domain fuzzy sets.

into a neural network. This helps the neural network in two ways. First, it does not need to learn from scratch, i.e., the neural network is already prestructured instead of consisting of random weights when the learning process starts. Secondly, if the data set is incomplete with respect to some 'corners' of the universe of discourse the fuzzy system might fill these gaps with expert know how. Such a combination of fuzzy systems and neural networks becomes even better if there is a way to translate the thus obtained neural network back into a fuzzy system. This means that the neural network is not anymore a black box that does not have any semantical meaning but can be interpreted by a fuzzy system (cf. Fig 6). Siemens uses this approach in several projects.

Genetic algorithms are like neural networks appropriate to learn input-output relations from data sets. Genetic algorithms have the advantage that they can 'jump' through the input-output domain while searching for local and global minima and maxima. So, genetic algorithms can play the role of the neural networks in Figs. 3b and c, although usually genetic algorithms are to slow to be used in online processes like presented in item c. One disadvantage of the methods b and c described above is that usually many input fuzzy sets are generated during the learning process. One can imagine that more than, say, 10 fuzzy sets on one domain that are partially overlapping each other may lead to rule bases that are hard to interpret. Genetic algorithms have been used to

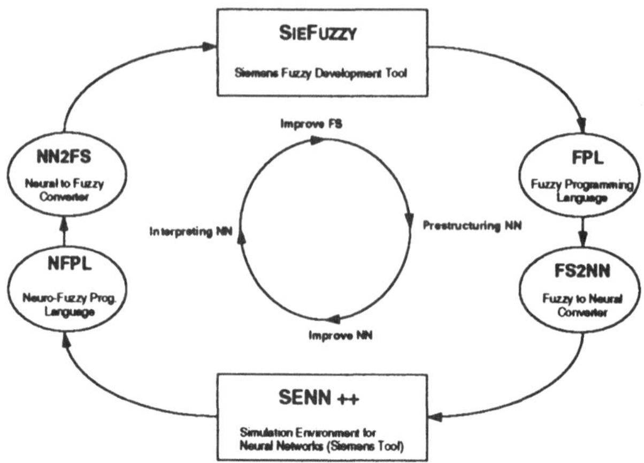

Figure 6: A combination of neural networks and fuzzy systems. The fuzzy system can be translated to a neural network, then the neural network can learn from data. The result can be translated back to a fuzzy system.

eliminate some of these fuzzy sets with only a small predefined loss of quality.

The future role of fuzzy systems in combination with chaos theory or nonlinear dynamics is hard to forecast. Goldstar in Korea does not provide much information about their 'fuzzy-neural-chaos' washing machine, but it seems that chaos theory is used in the simulation environment of the process to cause 'unexpected' experiences such that testing in the simulation environment can make more sense.

4 Design of Fuzzy Systems

In [Hellendoorn93] we have extensively described the design process as sketched in Fig. 7. We will now only summerize the design questions and then say more about practical design problems.

The first question that arises when one has to solve a particular problem is: "*Should this problem be solved with fuzzy control?*" The more general question that has to be answered is "Which classes of problems

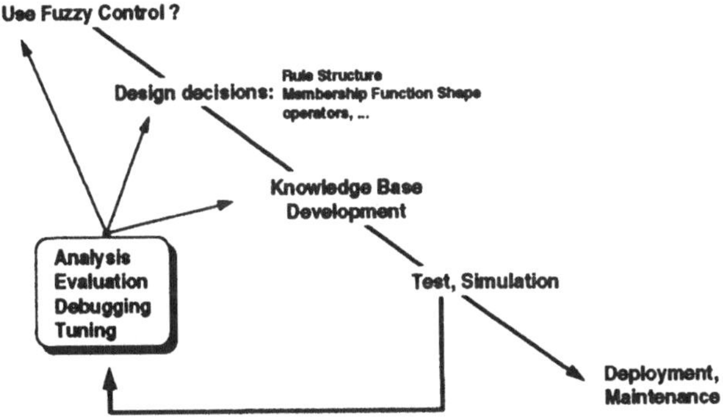

Figure 7: The design and development process of a fuzzy controller.

can be solved with fuzzy control?" The second question is: "*Which design parameters should be used for this problem?*" We expect that in the future we can figure out a number of certain classes, which describe various basic types of problems. So, the more general question that has to be answered is: "To which class of problems does this problem belong?" The third question is: "*What are the fuzzy rules for this problem?*" or, alternatively, "How does the rulebase look like?" This question can be answered with the help of experience in the development of conventional expert systems, or by using neural networks or genetic algorithms to learn the fuzzy system. The fourth question is: "*Which hardware platform has to be used?*" The answer to this question is highly problem dependent. Possible decision criteria are: real-time requirements, resolution, price, standardization, *etc.* The fifth question is: "*How should a fuzzy controller be tested?*" We distinguish between three different types of testing respectively analyzing: (1) Static testing: This means that the fuzzy controller is not integrated in any system but the controller itself is analyzed. (2) Closed-loop simulation: This method assumes that a model of the process to be controlled is available. (3) Closed-loop online analysis: This method avoids the modeling of the process to be controlled but requires advanced tools which allow graphical online-tuning (modification) and online debugging (representation) via communication with the hardware platform. The sixth question that has to be answered: "*How should the process model be build?*" There is no general consensus how to build the model. Depending on the required preciseness and availability one can choose mathematical models, described, e.g., by differential equations as well as fuzzy models, described by linguistic rules. Also various process identification techniques may be appropriate to describe the principles of the process to be controlled [Driankov/Hellendoorn/Reinfrank93].

In the design process of a fuzzy controller several extra problems have to be solved. Here one has to compare the design process of fuzzy controllers with that of conventional controllers, where one has to deal with several complicated process properties, like abrupt changes of the setpoint. This may lead to a large control output. Therefore, in conventional control usually the setpoint changes are not computed by the controller. In fuzzy control there studies in the literature with respect to this derivative kick.

Another point is the combination of a fuzzy controller with a high-pass or low-pass filter to dampen specified frequencies. In conventional control these filters are integrated in the design of the control system. There are no studies in fuzzy control which show the feasibility of designing a combination of FLC-PI and FLC-PD controllers.

A big problem in the control of actual processes is the sampling time. In fact, finding the right sampling time still remains more of an art than science. One has make a trade-off between process dynamics and computer capacity at disposal. Furthermore, one has to keep in eye the disturbance characteristics. Many papers regarding to this subject in conventional control theory were published, but in the fuzzy literature this problem has seldomly been described. Time delay is an even more complicated problem. This problem occurs due to the presence of distance lags, recycle loops or dead time. The time delay makes information from the true process arrive later than desired for the controller. Generally speaking, all information that is too old causes problems. They limit the performance of the control system and may lead to system instability. In the case of fuzzy control only few studies to this phenomenon exist. Because of the limited number of parameters, a conventional PID controller can not arbitrarily influence a process with higher-order dynamics. In systems with significant oscillations a higher-order regulator is needed. A general PID controller gives the necessary freedom to adjust for complex dynamics, and the same is valid for its discrete version. For a n-th order process it is well known that the order of the controller must be $n-1$, and in this way we have a PID^{n-1} controller. In the case of fuzzy control there are relatively few attempts to design controllers that deal with higher order system dynamics. For example, in [Boverie et al.93] the FLC is designed as a parallel structure of $n - 1$ single-input/single-output fuzzy controllers augmented with an integrator.

As in the case of conventional controllers, we have two major types of fuzzy controllers: a position type fuzzy controller known as PD-FC, and a velocity type fuzzy controller known as PI-FC. There has been little work done comparing the performance of these two types of fuzzy controllers but some recent studies lead to the following conclusions [Lee/Chae93]:

> "The PI-FC gives better performance in steady state but performs rather poorly during the transient. To improve its performance during transient appears to be difficult especially for a higher order system. Even in the case of a second order system the maximum variation of the incremental control output has to be limited to a rather small one to reduce overshoot of the transient response. However, this is a cause for increase in rise time. Thus to define the maximum variation of the incremental control output which gives both satisfactory rise time and minimum overshoot is a difficult task."

One natural way to overcome the above situation is introduce the second derivative of error as an additional input to the fuzzy controller (the other two being error and change of error). However it is not easy to measure the instantaneous value of this quantity nor is it easy to define the boundaries of its universe of discourse over which its fuzzy values have to be defined.

We could proceed longer in enumerating problems that have to be solved to come to a mature design and development methodology (cf. [Driankov/Hellendoorn/Palm94]). Future trends in fuzzy control are going into a direction to get better understanding of the system parameters and their correspondence with the process to be controlled.

5 Outlook

In this paper we have shown some trends in fuzzy. Firstly, new application trends in fuzzy like fuzzy classification and diagnosis that are going to play an important role besides the traditional fuzzy control. Secondly, the combination of fuzzy and other modern technologies like neural networks, genetic algorithms and chaos theory. Thirdly, the development of a design methodology for fuzzy systems which includes a design strategy how to optimize a fuzzy controller and ways to deal with complex process properties.

The next four papers deal with several of these new trends. In *Input Scaling of Fuzzy Controllers* Rainer Palm (Siemens R&D) deals with the problem how to develop an optimally adjusted fuzzy controller. In conventional control theory there has been done much research in this direction. The problem in fuzzy control is that there many parameters that can be changed. It is usual to start the tuning process with the scaling factors, more or less similar to conventional control. Dr. Palm shows a way to scale the input domains with the help of correlation coefficients.

Fig. 7 shows the design and development process of fuzzy systems. One important aspect is the development of the knowledge base which includes knowledge acquisition from experts or machine learning techniques to derive rules, but also consistency and completeness checks. The next question then is how to store the rule base. There are usually two ways, the first is to use a look-up table and interpolation methods which saves memory but costs run time, because interpolation has to be performed during run time. The second way is to store the whole control surface which saves run time but costs a lot of memory, in particular in high dimensional input spaces (exponential). Dr. Klaus Dieter Meyer-Gramann shows a way out of this dilemma in his paper *How to Store Efficiently a Linguistic Rule Set in a Fuzzy Controller*.

Jörn Hopf and Frank Klawonn in their paper *Learning the Rule Base of a Fuzzy Controller by a Genetic Algorithm* also deal with the knowledge base. They present a method based on genetic algorithms to design the knowledge base of a fuzzy system. This method can be compared with Fig. 3b where neural networks were used to produce fuzzy rules although genetic algorithms demand more information, e.g., optimization criteria and a model of the process to test the quality of a newly produced fuzzy controller.

The paper by Heiko Knappe, *Comparison of Conventional and Fuzzy-Control of Non-Linear Systems*, shows in an excellent way the correspondence between design methods in conventional and fuzzy control illustrated by an example in multi-mass systems with special nonlinear disturbances. It shows the importance of concepts like speed control and position control in fuzzy systems (cf. [Driankov/Hellendoorn/Palm94]). Furthermore, it shows some principal properties of fuzzy controllers and many research problems.

References

[Bezdek81] J.C. Bezdek, *Pattern Recognition with Fuzzy Objective Function Algorithms*, New York, Plenum Press, 1981.

[Boverie et al.93] S. Boverie et al., "Performance Evaluation of Fuzzy Control through an International Benchmark," *Proceed. IFSA'93 World Congress,* pp. 941-944, Seoul, Korea, 1993.

[Driankov/Hellendoorn/Palm94] D. Driankov, R. Palm & H. Hellendoorn, "Some Research Directions in Fuzzy Control." In: H.T. Nguyen, M. Sugeno, R.R. Yager, *Theoretical Aspects of Fuzzy Control*, J. Wiley, 1994

[Driankov/Hellendoorn/Reinfrank93] D. Driankov, H. Hellendoorn & M. Reinfrank, *An Introduction to Fuzzy Control*, Heidelberg, Springer Verlag, 1993.

[Dubois/Prade80] D. Dubois & H. Prade, *Fuzzy Sets and Systems: Theory and Applications,* Orlando (Florida), Academic Press, 1980, 393p.

[Hellendoorn93] H. Hellendoorn, "Design and Development of Fuzzy Systems at Siemens R&D," *Proceedings of the IEEE International Conference on Fuzzy Systems*, San Francisco, 1993, pp. 1365–1370.

[Hellendoorn/Thomas93] H. Hellendoorn & C. Thomas, "On Defuzzification in Fuzzy Controllers," *Journal of Intelligent and Fuzzy Systems,* **1**(2)(1993)109–123.

[Lee/Chae93] J. Lee, & S. Chae, "A Fuzzy Logic Controller Using Error, Change-of-error, and Control Input," *Proceed. IFSA'93 World Congress,* pp. 956-959, Seoul, Korea, 1993.

[Osaki et al.] N. Osaki, K. Harayama, W. Shinohara & S. Hayashi, "Recovery Boiler Intelligent Control," *Proc. of the 3rd IFAC Int. Workshop on Artificial Intelligence in Real Time Control.*

[Palm94] R. Palm & D. Driankov, "Fuzzy Inputs," *Proc. of the FUZZ-IEEE'94*, Orlando, 1994

[Wang/Mendel91] L.-X. Wang & J.M. Mendel, "Back-Propagation Fuzzy System as Nonlinear Dynamic System Identifier," *IEEE Trans. on Neural Networks*, 1991

2.2. Input Scaling of Fuzzy Controllers

Rainer Palm

Abstract

The paper deals with the optimal adjustment of input scaling factors for Fuzzy Controllers. The method bases on the assumption that in the stationary case an optimally adjusted input scaling factor meets a specific statistical input output dependence. A measure for the strength of statistical dependence is the *correlation coefficient*. The article deals with the so-called equivalent gain which is closely connected to the cross-correlation of the controller input and the defuzzified controller output. Without loss of generality, the adjustment of input scaling factors using correlation functions is pointed out by means of a single input - single output (SISO) - system. The contribution concludes with a set of fuzzy rules controlling a redundant robot manipulator.

1 Introduction

Most fuzzy controllers (FC) work in such a way that, on the basis of *crisp desired inputs* and *crisp actual outputs*, the system (plant) is controlled also by *crisp manipulated variables*. In this wide-spread case *fuzziness* is restricted only to the controller which is known to be more robust than conventional controllers [Lee 90, Driankov 93, Palm 92b]. The operation of a FC of this type requires fuzzification of the inputs (e.g. error and change of error): each crisp input is attached to a subset of grades of membership depending on the *a priori* chosen subset of membership functions. The design of a FC requires information about the system to be controlled such as operating areas of the FC inputs and the manipulated variable of the FC output. For simplicity, in most cases the membership functions of the input and output variables are defined within normalized intervals (universes of discourse). In the case of normalized universes of discourse an appropriate choice of specific operating areas requires respective scaling or denormalization factors. An input scaling factor transformes a physical signal into the normalized universe of discourse of the controller input whereas the output denormalization factor provides a transformation of the defuzzified output signal from the normalized universe of discourse of the controller output into a physical manipulated variable. The importance of an optimal choice of input scaling factors is evidentely shown by the fact that ill scaling results in either shifting the operating area to the boundaries or utilizing only a small area of the normalized universe of discourse. On the other hand, the adjustment of the output denormalization factor affects the closed loop gain which has direct influence on system stability. The behavior of the system controlled finally depends on the choice of the normalized transfer characteristics (control surface) of the controller. In the case of a predefined set of rules the control surface is mainly determined by shape and location of the input and output membership functions. Taking these influences into account for controller design one should pay attention to the following priority list:

1. The output denormalization factor has the most influence on stability and oszillation tendency. Because of its strongest impact on stability this factor is assigned to the first priority in the design process.

2. Input scaling factors have the most influence on basic sensitivity of the controller with respect to optimal choice of the operating areas of the input signals. Therefore, input scaling factors are assigned to the second priority.

3. Shape and location of input and output membership functions and, with this, the transfer function of the normalized controller influence positively the behavior of the system controlled in different areas of the state space provided that the operating areas of the signals are optimally chosen through a well adjusted input scaling factor. Therefore, this aspect is getting the third priority.

Once, by means of some system analysis, the scaling factors and the parameters of the membership functions have been chosen the next task is to tune them in order to improve the systems's behavior according to some optimization criteria. In this context tuning should consider the same priority list as for the design process.

Most of the reports on tuning refer to membership functions in order to change the transfer characteristics of the controller [Smith 92, Zheng 93, Boscolo 92]. Examples for gain tuning can be found at [Katayama 91, Maeda 92, Viljamaa 93, Braae 79, Zheng 92]. Tuning of rules has been considered by [Procyk 79, Peng 90]. Many reports deal with integral criteria with respect to particular test signals such as step, pulse, and random functions, respectively [Smith 92, Zheng 93]. This work deals with the 2nd level of tuning hierarchy, namely with appropriate scaling of controller inputs which has the most influence on the sensitivity of the controller. It is assumed that both the rule set and the membership functions are predefined and kept constant during the tuning process. The input data are assumed to be statistically distributed satisfying a Gaussian distribution whose parameters are assumed to be unknown. A poor knowledge about the distribution parameters can be explained by slow varying plant parameters or drift in the sensory used for state observation. Here, we distinguish three relevant cases:

1. Known mean (e.g. mean=0) and unknown deviation.
2. Unknown mean and known deviation.
3. Mean and deviation are unknown.

Considering a controller with multi input and single output cases 1 - 3 can be processed on-line measuring the linear dependence between each input and output signal of the controller. A measure for linear input-output dependence of a transfer element is the cross-correlation function and the cross-correlation coefficient, respectively. *Firstly*, the shift of the signal's mean value along the universe of discourse ensures the signal to meet the relevant operating area of the control surface. *Secondly*, once the relevant region is reached the tuning procedure keeps on changing the particular input scaling factor until the goal, the cross-correlation coefficient to be a certain value near its maximum, is reached. It is shown that for Gaussian input signals a given FC can be imaginarily replaced by an *equivalent gain* which strongly depends on the nonlinear transfer characteristic of the FC [Schlitt 68]. This method allows the utilization of linear system theory even in the case of nonlinear elements within the control loop. Therefore, an appropriate choice of the equivalent gain has a great influence on the behavior of the closed loop system. The equivalent gain can be expressed in terms of the standard deviation of the input and the *input-output cross-correlation function*. The claim is that for a stationary input a certain amount of signal amplitudes around the operating point of the FC should be linearily transmitted by the FC.

As already mentioned, a measure for *linear* input-output dependence of a transfer element is the cross-correlation function. Hence, if a specific linearity between input and output is required one has to adjust the standard deviation in such a way that a corresponding cross-correlation coefficient is met. For a given SISO FC structure the only parameter to influence the equivalent gain is the scaling factor for the input signal. This result can easily be extended to the Multi Input/Multi Output case (MIMO) if the individual states of the plant to be controlled are non-correlated with each other. The method presented deals with the optimal adjustment of scaling factors for FCs with the help of the input-output cross-correlation [Palm 93]. If the distribution of the signal is a *priori* known the method can be characterized as a design approach only by consideration of the nonlinear FC without closing the control loop. An example shows how to tune scaling factors in the process of kinematical control of a redundant robot manipulator.

2 Input output correlation for a FC

Equivalent gain - SISO case Fuzzy controllers with crisp inputs and outputs can be considered as multidimensional nonlinear transfer elements with upper and lower limits whose transfer characteristics at the origin can be described by

$$\bar{u}(0) = 0. \tag{1}$$

This can be justified by the following reasons:

1. Most important applications deal with the error e or error vector $\mathbf{e} = (e, \dot{e}, \ldots)^T$ at the controller's input so that the respective input signal is centralized.
2. A stationary signal can, with the knowledge of its mean value, always be centered so that a centralized signal at the controller input can be assumed.

Let the system to be controlled be linear or, within the operating area, linearizable with lowpass characteristic (see fig. 1). Furthermore, let the desired value w include Gaussian noise. We then obtain non-Gaussian noise at the output of the FC because of its nonlinear transfer characteristic. On the other hand, we suppose the system to filter out all frequencies causing a non-Gaussian distribution. In this way, we expect to have Gaussian noise at the output of the system and, with this, at the adder where the desired value w is compared with the actual value x. With these assumptions the scaled signal $e_s = (w - x) \cdot s_c$ is also of Gaussian type. From nonlinear system theory we know the ***describing function*** for sinusoidal signals and the ***equivalent gain*** for signals with noise [Schlitt 68]. The basic idea of this method is to substitute imaginarily the nonlinear element in a closed loop system by a linear one whose gain depends on the amplitude e_0 (for sinusoidal inputs) or variance σ_e^2 (for noise) of the controller input. The main purpose of this approach was the stability test of the nonlinear system to be controlled with the means of linear control theory. This method is here adopted for the adjustment of scalings of a given FC.

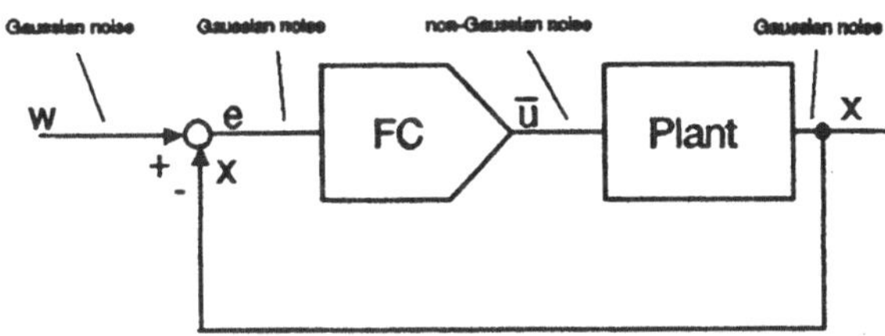

Figure 1: Closed loop system with a nonlinear FC

Let e be a stationary and ergodic Gaussian process.
Furthermore, let

$$\bar{u} = K(\sigma_e^2) \cdot e + v \tag{2}$$

where

$K(\sigma_e^2)$ - equivalent gain corresponding to a specific FC transfer characteristic
v - noise which is non-correlated with e_s.

Let, finally,

$$R_{e,\bar{u}}(\sigma_e) = E[(e(t) - E[e(t)]) \cdot (\bar{u}(t) - E[\bar{u}(t)])] \tag{3}$$

be the linear cross-correlation function for $\tau = 0$, and

$$E[x(t)] = \lim_{t \to \infty} \frac{1}{2T} \int_{-T}^{+T} x(t)dt.$$

the expected value.

For simplicity the mean values of e, $\bar{u}$, and v are equal to zero:

$$E[e] = 0; \qquad E[\bar{u}] = 0; \qquad E[v] = 0.$$

Multiplying both sides of eq.(2) with e and computing the expected values one obtains

$$E[e \cdot \bar{u}] = K(\sigma_e^2) \cdot E[e^2] + E[e \cdot v] \tag{4}$$

with

$E[e \cdot \bar{u}]$ - cross-correlation $R_{e\bar{u}}$
$E[e^2]$ - variance σ_e^2.

Because of $E[e \cdot v] = 0$ one obtains for the equivalent gain

$$K(\sigma_e^2) = \frac{R_{e\bar{u}}(\sigma_e)}{\sigma_e^2}. \tag{5}$$

With regard to scaled Gaussian distributed input signals e_s we then obtain the equivalent gain

$$K(\sigma_{e_s}) = \frac{R_{e,\bar{u}}(\sigma_{e_s})}{\sigma_{e_s}^2}. \tag{6}$$

Input scaling The input e is scaled by means of

$$e_s = e \cdot s_c \tag{7}$$

where

$$\begin{aligned} s_c &\quad - \quad \text{scaling factor,} \\ e_s &\quad - \quad \text{scaled input.} \end{aligned} \tag{8}$$

Let the scalar output signal $\bar{u}$ be computed by the center of gravity:

$$\bar{u} = \int_A^B \frac{\mu_u \cdot u}{\int_A^B \mu_u du} du \tag{9}$$

where

$$\begin{aligned} \mu_u \in (0,1) \quad &- \quad \text{degree of membership,} \\ u \in (A, B) \quad &- \quad \text{universe of discourse.} \end{aligned} \tag{10}$$

For simplicity we assume the denormalization factor of $\bar{u}$ to be one. Furthermore, let $e(t)$ and $\bar{u}(t)$ be stationary and ergodic processes. The standard deviations of the scaled signal $e_s(t)$ and non-scaled signal $e(t)$ are connected in the same way as the signals e_s and e are:

$$\sigma_{e_s} = s_c \cdot \sigma_e. \tag{11}$$

From (6) and (11) we get

$$K(\sigma_e) = \frac{R_{e,\bar{u}}(s_c \cdot \sigma_e)}{s_c^2 \cdot \sigma_e^2}. \tag{12}$$

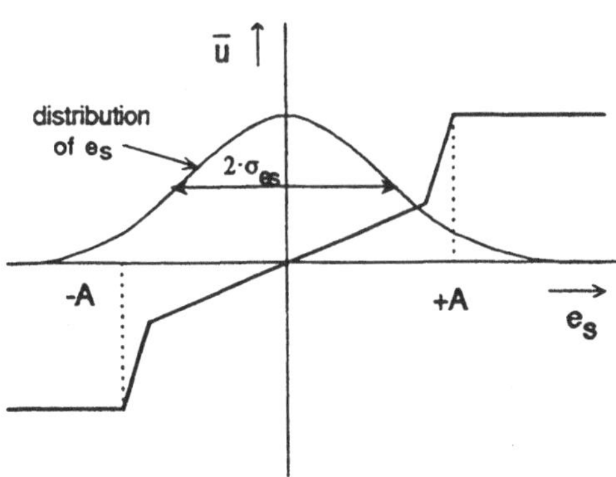

Figure 2: Symmetrical nonlinear transfer characteristic of a FC with limits

Both gain K and cross-correlation $R_{e,\bar{u}}$ reach their maximum values in the case of maximum linear input-output connection. The normalized correlation coefficient

$$\tilde{R}_{e,\bar{u}} = \frac{R_{e,\bar{u}}}{\sigma_{e_s} \cdot \sigma_{\bar{u}}} \tag{13}$$

reaches its maximum at $\tilde{R}_{e,\bar{u}}|_{max} = 1$. $\tilde{R}_{e,\bar{u}}$ reaches its minimum at

$$\tilde{R}_{e,\bar{u}}|_{min} = \sqrt{\frac{2}{\pi}}$$

if the transfer characteristic is symmetrical with respect to the mean value $\bar{e}_s = E[e_s(t)]$ (see fig.2). This is related to a relay transfer characteristic because for a very large standard deviation compared with the width $2 \cdot A$ of the interval considered every symmetrical transfer characteristic acts as a relay function. Moreover, it is evident that a shift of the mean $\bar{e}_s$ of the distribution to the limits of the transfer characteristic of the FC leads to decreasing input-output correlation of the controller. The extreme point is reached when the largest part of the distribution is covered by one of the branches at which the control output is always a constant value. For this case we obtain $\tilde{R}_{e,\bar{u}} \approx 0$.

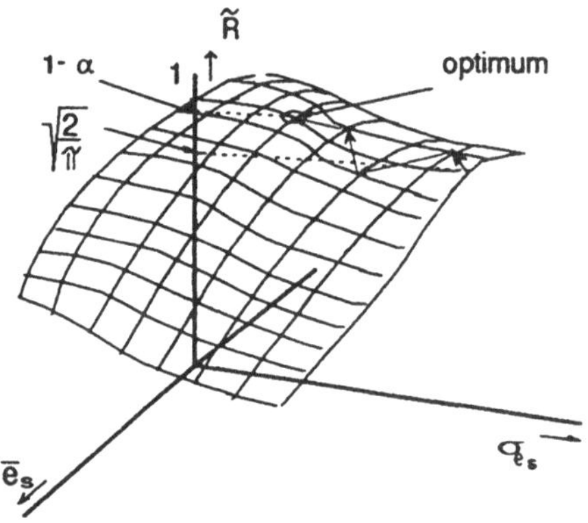

Figure 3: $\tilde{R}(\bar{e}_s, \sigma_{e_s})$-plot

Optimal input scaling actually means: searching for an optimal σ_{e_s} with respect to the interval [-A,+A] of the FC between its limits. We assume the optimal scaling for the following case:
We start the searching procedure with a large s_c (which means a large $\sigma_{e_s} = \sigma_{e_s}|_{max}$). Then, keeping σ_{e_s} constant, we change $\bar{e}_s$ stepwise by adding $\Delta\bar{e}_s$ to $\bar{e}_s$ which corresponds to a shifting of the distribution of the input signal along the $\bar{e}_s$-axis. The result is a curve $\tilde{R}(\bar{e}_s, \sigma_{e_s})$ with a single maximum $\tilde{R}(\bar{e}_s, \sigma_{e_s})|_{max}$ with

$$\tilde{R}(\bar{e}_s + \Delta\bar{e}_s, \sigma_{e_s}) \neq \tilde{R}(\bar{e}_s, \sigma_{e_s}) \qquad \forall \tilde{R}(\bar{e}_s, \sigma_{e_s}). \tag{14}$$

and

$$\sigma_{e_s} = const.$$

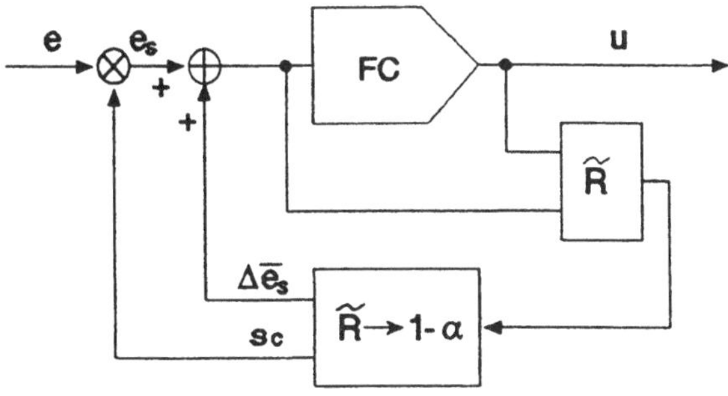

Figure 4: Block scheme for tuning of scaling factors

After that we decrease σ_{e_s} by $\Delta\sigma_{e_s}$ and change the result $\tilde{R}(\bar{e}_s, \sigma_{e_s} - \Delta\sigma_{e_s})$ in the same way. Because of the monotonic function $\tilde{R}(\bar{e}_s, \sigma_{e_s})|_{\bar{e}_s=const.}$ we obtain a higher maximum $\tilde{R}(\bar{e}_s, \sigma_{e_s})|_{max}$ as before:

$$\tilde{R}(\bar{e}_s, \sigma_{e_s} - \Delta\sigma_{e_s})_{max} > \tilde{R}(\bar{e}_s, \sigma_{e_s})_{max}. \tag{15}$$

We stop the searching procedure at

$$\tilde{R}(\bar{e}_s, \sigma_{e_s}) \geq 1 - \alpha$$

with condition (14) where $\alpha \in (0,1)$ is a free parameter. If condition (14) is not fulfilled we obtain a plateau. In this case the domain of the FC is assumed to be insufficient with respect to the given standard deviation σ_{e_s}. Hence, one has to increase the scaling factor s_c until (14) is met. Figure 3 shows a typical $\tilde{R}(\bar{e}_s, \sigma_{e_s})$-plot and fig. 4 shows the corresponding block scheme.
We choose the free parameter α such that for a linear FC characteristic between the upper and lower limit the standard deviation σ_{e_s} of the scaled signal e_s is set equal to the interval A of the controller:

$$\alpha = 1 - \tilde{R}(\bar{e}_s, \sigma_{e_s})_{\sigma_{e_s}=A} \tag{16}$$

This means, we have an input signal probability $P = 0.68$ for the linear region of the FC. However, if the characteristic of the controller between its limits is nonlinear (see fig. 2) one obtains automatically an s_c such that $\sigma_{e_s} < A$. Namely, it is clear that the cross-correlation $\tilde{R}$ as a linear operation meets its maximum value when, for a given standard deviation σ_{e_s}, the function between the limits of the transfer characteristic of the FC is linear. If the function between the limits is nonlinear one obtains for the same σ_{e_s} a lower value for $\tilde{R}$. This, however, corresponds to a smaller standard deviation σ_{e_s} and, with this, a smaller scaling factor s_c if a linear function between the limits is assumed.

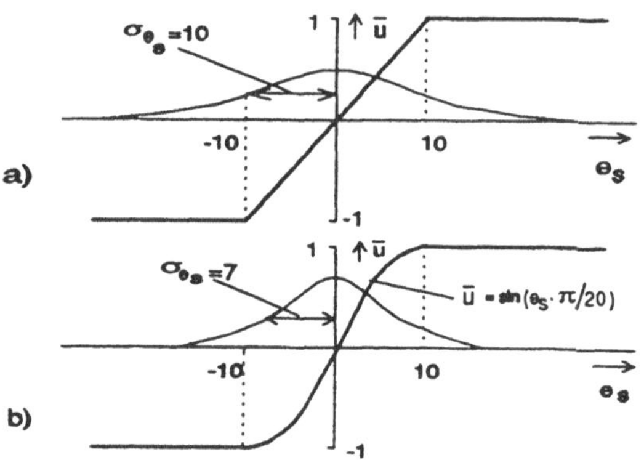

Figure 5: FCs with linear (a) and nonlinear (b) characteristics between the lower and upper limits

Numerical example Suppose a standard deviation $\sigma_e = 1$ of the input signal $e(t)$. For a FC with a linear characteristic between the limits as shown in fig. 5a) we obtain $\tilde{R}(\bar{e}_s, \sigma_{e_s})_{\sigma_{e_s}=10} = 0.95$ and $\alpha = 0.05$. This corresponds to a scaling factor $s_c = 10$ and the scaled standard deviation $\sigma_{e_s} = 10$. For a FC with a sinusoidal characteristic as shown in fig. 5b) we obtain for the same $\alpha = 0.05$ a scaling factor $s_c = 7$ and the scaled standard deviation $\sigma_{e_s} = 7$.

3 Application to a redundant manipulator arm

The application sample of this section deals with the control of the kinematic of a redundant robot arm (the inverse task) whose static transfer

characteristic is highly nonlinear but, in contrast to this, whose internal dynamics consists of a simple dead time element or a delay. Contents of this section is, however, not to describe the whole problem of solving the inverse task in the case of kinematical redundancy. This has already been discussed at [Palm 92a] in detail. In the following, only the aspect of appropriate tuning of the input scaling factors required for kinematical control of the robot arm is considered.
In section 2 it has been shown that a shift of the mean value of an input signal leads to a reduction of the correlation coefficent. The following example deals with input signals whose signs do not change during the control process. Moreover, the input scaling factor does not only affect the standard deviation but the mean value, too. The task is the optimal choice of scaling factors so that the distribution of the corresponding input signal is located within the corresponding operating area of the FC. The basic assumption is that an optimal scaling factor is obtained when the statistical input-output dependence meets its maximum:

$$|\tilde{R}_{\tau=0}| \longrightarrow max.$$

To be independent of any change of sign in the control loop the absolute value of $\tilde{R}$ has been chosen. The problem of kinematical control of a redundant manipulator arm can be simply described as follows: The effector (gripper, tool) of the planar robot arm is supposed to follow a predefined path (see fig. 6). The robot kinematic is constructed in such a way that the individual links of the manipulator are able to avoid both external obstacles and internal restrictions (e.g. boundaries of the links).

In the special case the motion of each link is, in addition to the given effector task, determined by

1. distance h between link and middle position,
2. distance s between link and wall,
3. distance d between link and obstacle.

The distances h, s and d are evaluated by fuzzy attributes (e.g. s=Positive Small) and their membership functions. For each link a fuzzy rule base provides the corresponding correction of the joint angle. The actions z (angle corrections) of each link are evaluated also by fuzzy attributes (e.g. z=Positive Big angle correction). Although the process to be controlled is highly nonlinear in the large it can be considered as linear for small angle corrections. The distances h, s and d are scaled so that they fit the predefined normalized universes of discourse.

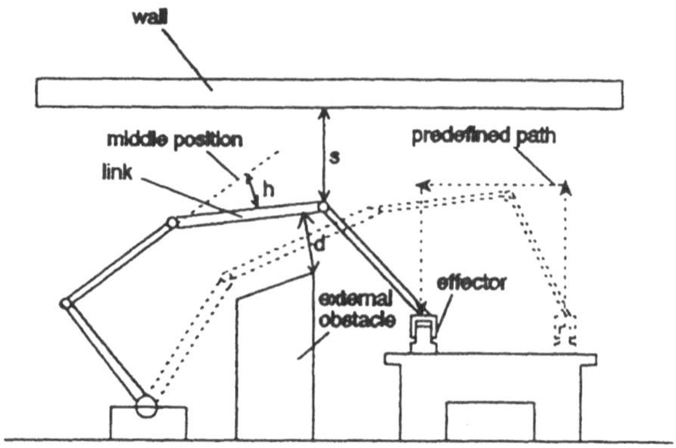

Figure 6: Motion of a redundant robot arm

For the internal restriction "Scaled distance between the i-th link and its middle position" $h_{iN} = h$ yields

$$\begin{aligned} \text{h negative big:} \quad & \text{HNB}=(\mu_{HNB}(h)/h), \\ \text{h negative small:} \quad & \text{HNS}=(\mu_{HNS}(h)/h), \\ \text{h positive big:} \quad & \text{HPB}=(\mu_{HPB}(h)/h), \\ \text{h positive small:} \quad & \text{HPS}=(\mu_{HPS}(h)/h), \\ & \forall h \in H. \end{aligned}$$

For the external restriction "Scaled distance between the i-th link and some wall" $s_{iN} = s$ yields

$$\begin{aligned} \text{s big:} \quad & SIB = (\mu_{SIB}(s)/s), \\ \text{s small:} \quad & SIS = (\mu_{SIS}(s)/s), \\ & \forall s \in S. \end{aligned}$$

For the external restriction "Scaled distance between the i-th link and some obstacle" $d_{iN} = d$ yields

$$\begin{aligned} \text{d big:} \quad & DIB = (\mu_{DIB}(s)/s), \\ \text{d small:} \quad & DIS = (\mu_{DIS}(s)/s), \\ & \forall d \in D. \end{aligned}$$

For the "Scaled output" $z_i = z$ according to the i-th link yields

$$\begin{aligned}
\text{z negative big:} &\quad \text{ZNB}=(\mu_{ZNB}(h)/h),\\
\text{z negative small:} &\quad \text{ZNS}=(\mu_{ZNS}(h)/h),\\
\text{z negative zero:} &\quad \text{ZNZ}=(\mu_{ZNZ}(h)/h),\\
\text{z positive big:} &\quad \text{ZPB}=(\mu_{ZPB}(h)/h),\\
\text{z positive small:} &\quad \text{ZPS}=(\mu_{ZPS}(h)/h),\\
\text{z positive zero:} &\quad \text{ZPZ}=(\mu_{ZPZ}(h)/h),\\
&\quad \forall z \in Z.
\end{aligned}$$

All membership functions μ vary only within a predefined standard interval $h, s, d, z \in [MAX, MIN]$. Outside this interval the value of μ is either 0 or 1. Furthermore, all fuzzy sets are normal, i.e. there is an h, s, d or z with a corresponding $\mu = 1$.

To obtain an appropriate motion for each link the following set of rules has been applied:

IF (SIS AND DIS AND (HNS OR HPS)) OR (SIS AND HNB AND DIB) THEN ZNZ

IF (SIS AND HPB AND DIS) OR (SIB AND HPS AND DIB) THEN ZNS

IF (SIS AND DIB AND (HNS OR HPS)) OR ((SIS OR SIB) AND HPB AND DIB) THEN ZNB

IF SIS AND HNB AND DIS THEN ZPZ

IF (SIB AND DIS AND (HPS OR HPB)) OR (SIB AND HNS AND DIB) THEN ZPS

IF (SIB AND DIS AND (HNS OR HNB)) OR (SIB AND HNB AND DIB) THEN ZPB

The scalar output value has been computed by the center of gravity:

$$z_N = \frac{\int_{z_{min}}^{z_{max}} z \cdot \mu_z dz}{\int_{z_{min}}^{z_{max}} \mu_z dz}.$$

Within the rules for the operations AND and OR the MAX and MIN operator, respectively, have been chosen.

The correlation coefficient $\tilde{R}$ for discrete points of time has been applied concerning the distance s between each link and the wall:

$$\tilde{R}[s,z] = \frac{\sum_{i=1}^{n} s_i z_i - \frac{1}{n}(\sum_{i=1}^{n} s_i \cdot \sum_{i=1}^{n} z_i)}{\sqrt{\sum_{i=1}^{n} s_i{}^2 - \frac{1}{n}(\sum_{i=1}^{n} s_i)^2} \cdot \sqrt{\sum_{i=1}^{n} z_i{}^2 - \frac{1}{n}(\sum_{i=1}^{n} z_i)^2}}. \tag{17}$$

Figures 7 and 8 show the change of the correlation coefficient $\tilde{R}$ where s_s is the scaling factor for distance s.

The other scaling factors are $s_h = 20$ and $s_d = 120$ (see fig. 7). The peak of $\tilde{R}$ is lying at about $s_s = 80$. Figure 8 shows a similar situation for $s_h = 100$ and $s_d = 60$. The peak of $\tilde{R}$ lies at $s_s = 80$ again. The result is: Although there is a certain change in the curvature of $\tilde{R}(s_s)$, depending on a different choice of the other scaling factors s_d and s_h, the abscissa of the maximum value of $\tilde{R}$ does not change. This finally illustrates the independence of the location of the maxima of different correlation coefficient curves.

4 Conclusion

Many control applications show that most fuzzy controllers are designed in such a way that the universes of discourse concerning the membership functions used are normalized according to a standard interval. This leads to the task of an appropriate choice of scaling factors for inputs and outputs. Together with an appropriate adjustment of membership functions Input/output scaling forms a tuning hierarchy in which input scaling gets the second priority after tuning the output scalings and before tuning the membership functions. Optimal adjustment of input scaling factors serves as a mean to influence the basic sensitivity of the controller and is the basis for the relevant utilization of the operating areas of the input signals. The basis of the method is a well-known approach of the nonlinear control theory where under certain conditions for Gauss-distributed input signals a nonlinear control element can be imaginarily replaced by the so-called equivalent gain which is closely related to the input-output cross-correlation coefficient of the controller. The claim of the method presented is that for a stationary input signal a certain amount of its amplitudes should be linearily transmitted through the FC. Under the condition that the system to be controlled is approximately a lowpass filter one is able to adjust the input scaling factors at run-time measuring the cross-correlation coefficient. Another case comes up if the distribution of the regarding input signal is known. In this case a tuning procedure becomes pointless and one is therefore able to design the input scaling factors from the beginning so that the conroller meets its input sensitivity required. Once, however, either mean or deviation of the input signal is unknown tuning with respect to the procedure described is justified.

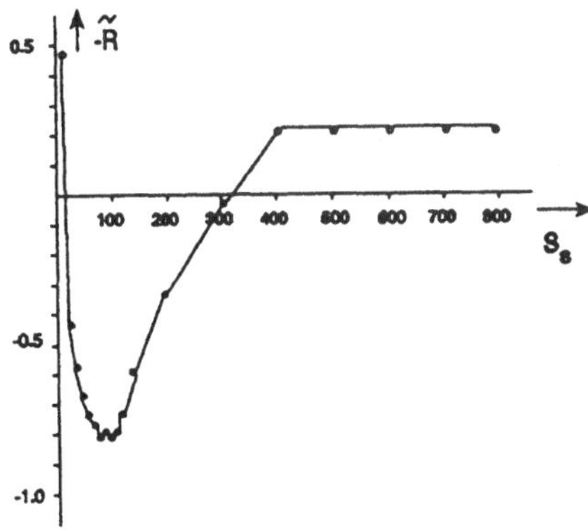

Figure 7: Simulation results for s_s with $s_h = 20$ and $s_d = 120$

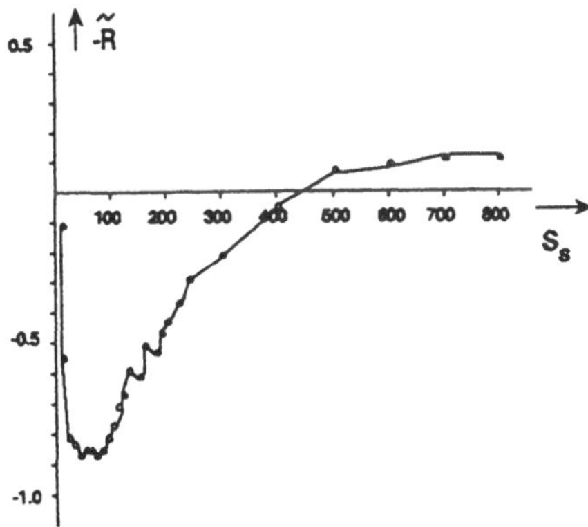

Figure 8: Simulation results for s_s with $s_h = 100$ and $s_d = 60$

In an example it is shown that the kinematical configuration of a redundant robot arm is controlled by means of a set of fuzzy rules. This example shows the applicability of the correlation method for the adjustment of input scaling factors in the MISO-case even if the system to be controlled is highly nonlinear.

References

[Boscolo 92] A.Boscolo, F.Drius: Computer Aided Tuning and Validation of Fuzzy Systems.*1st IEEE Int. Conf. on Fuzzy Systems 1992* San Diego March 1992, pp. 605-614

[Braae 79] M.Braae, D.A. Rutherford: Selection of Parameters for a Fuzzy Logic Controller. *Fuzzy Sets and Systems 2 (1979), North-Holland* pp.185-199

[Driankov 93] D.Driankov, H.Hellendoorn, M.Reinfrank: An Introduction to Fuzzy Control. *Springer Berlin Heidelberg New York 1993*

[Gibson 63] J.Gibson: Non-linear Automatic Control *McGraw-Hill, New York* 1963

[Katayama 91] R.Katayama, Y.Kajitani, K.Matsumotu, M. Watanabe, Y.Nishida: An Automatic Knowledge Acquisition and Fast Self Tuning Method for Fuzzy Controllers Based on Hierarchical Fuzzy Inference Mechanisms. *IFSA '91 Brussels July 7-12* Proceedings Volume AI pp. 105-108

[Lee 90] Chuen Chien Lee: Fuzzy Logic in Control Systems: Fuzzy Logic Controller-Part I. *IEEE Trans. on Syst. Man, and Cyb., Vol. 20. No. 2. March/April 1990* pp.404-418

[Maeda 92] M.Maeda, S.Murakami: A self-tuning fuzzy controller. *Fuzzy Sets and Systems 51 (1992) North-Holland* pp. 29-40

[Palm 92a] R.Palm: Control of a Redundant Manipulator Using Fuzzy Rules. *Fuzzy Sets and Systems (1992), Vol. 45 No. 3, North-Holland* pp.279-298

[Palm 92b] R.Palm: Sliding Mode Fuzzy Control, *1st IEEE International Conference on Fuzzy Systems 1992, Fuzz-IEEE'92-Proceedings* San Diego March 8-12, pp.519-526

[Palm 93] R.Palm: Tuning of Scaling Factors in Fuzzy Controllers Using Correlation Functions.*2nd IEEE International Conference on Fuzzy Systems 1993* San Francisco, March 1993, pp. 279-298

[Palm 94a] R.Palm, M.Reinfrank, K.Storjohann: Towards Second Generation Applications of Fuzzy Control. *12th IFAC World Congress 1994, July 18-23 1993 Sydney*proceedings vol.5 pp.549-552

[Palm 94b] R.Palm, D. Driankov: Fuzzy Inputs. *FUZZ IEEE'94, Orlando June 26-29, 1994* to appear.

[Driankov 94] D.Driankov, R.Palm, H.Hellendoorn: Fuzzy Control With Fuzzy Inputs: The Need for New Rule Semantics. *FUZZ IEEE'94, Orlando June 26-29, 1994* to appear.

[Peng 90] Peng Xian-Tu: Generating Rules for Fuzzy Logic Controllers by Functions. *Fuzzy Sets and Systems 36 (1990), North-Holland* pp.83-89

[Procyk 79] T.J. Procyk, E.H. Mamdani: A Linguistic Self-Organizing Process Controller. *Automatica Vol. 15 (1979), Pergamon Press Ltd.* pp. 15-30

[Schlitt 68] H.Schlitt: Stochastische Vorgaenge in linearen und nichtlinearen Regelkreisen. 1968 by Vieweg, Braunschweig

[Smith 92] S.M.Smith, D.J.Comer: An Algorithm fo Automated Fuzzy Logic Controller Tuning. *1st IEEE International Conference on Fuzzy Systems 1992, Fuzz-IEEE'92-Proceedings* San Diego March 8-12, pp.615-622

[Viljamaa 93] P.Viljamaa, H.K.Koivo: Tuning of Multivariable Fuzzy Logic Controller. *2nd IEEE International Conference on Fuzzy Systems 1993* San Francisco, March 1993, pp. 697-701

[Zheng 92] Li Zheng: A Practical Guide to Tune PI-Like Fuzzy-Controllers *1st IEEE International Conference on Fuzzy Systems 1992, Fuzz-IEEE'92-Proceedings* San Diego, March 8-12, pp.633-640

[Zheng 93] Li Zheng: A Practical Computer-Aided Tuning Technique for Fuzzy Control. *2nd IEEE International Conference on Fuzzy Systems 1993* San Francisco, March 1993, pp. 702-707

2.3. How to Store Efficiently a Linguistic Rule Set in a Fuzzy Controller

K.-D. Meyer-Gramann

Abstract

A fuzzy controller can determine its output values either by evaluating a fuzzy look-up table or by interpreting directly its linguistic rule set. Sometimes the second way has advantages. In this case a method has to be derived to store the rule set in the actual runtime environment such that the controller can interprete it sufficiently fast at run-time.

In this article a fundamental solution of this task is presented. The original rule set is transformed by replacing rules with an identical conclusion by one single rule. If necessary, the premise of such an integrated rule is reformulated such that each rule premise has a standard structure - an OR-connection of AND-connections of linguistic values. A standard structure premise is coded as a matrix with integer-valued elements. It is shown in detail how to code a premise and what are the storage requirements of this method.

It is demonstrated how the controller calculates its output values during run-time. For each sample point it determines the truth value of each linguistic rule premise by evaluating its corresponding premise matrix.

Fuzzy control was implemented with the help of standard functional units of an automatization language. This approach integrate the novel technique of fuzzy control into an automatization environment. As an application of the introduced storing method, it is shown how one can realize the presented approach by additional functional units.

1 Introduction

How to implement a fuzzy controller?

The most fuzzy controllers are implemented as pure software programs of a high-level language like C or PASCAL. The controller design tools being available on the market typically yield source code of such a programming language. The software solution is very flexible which is advantageous when the controller has to be changed. But it may require too much memory space or may run too slow. As an alternative solution, one can apply dedicated fuzzy hardware in order to obtain a very fast controller. But this specific hardware solution can often hardly be integrated into the existing control environment. In addition, it may be difficult to configure and maintain such a controller. One may depend on a single hardware vendor.

A third way to implement a fuzzy controller is to use standard functional units of a common automatization language. The engineer configures the controller with the help of a given set of standard functional units which is offered by the engineering tool. The set contains units for the "fuzzification", the fuzzy inference engine, and the defuzzification as well as data units for storing the linguistic rules and the membership functions.

No matter what method of implementing a fuzzy controller is chosen - one has to find a way which enables the fuzzy controller to perform the run-time evaluation of the linguistic rules. Two different approaches are suggested in the literature and are realized by commercially available engineering tools. The approaches describe how a fuzzy controller calculates its output value(s) as a function of its input value(s) during run-time:

- o The fuzzy controller evaluates the linguistic rule set directly.
- o One exploits the property that the behavior of a fuzzy controller is uniquely determined by its control surface: An output value for time t only depends on the input values for t but not on any previous input value or any other output value. A fuzzy controller itself has no memory and no feedback capability. When having designed and validated the controller, one creates a so-called "fuzzy look-up table" which

determines for each characteristic input value vector the corresponding output value vector. At run-time the controller evaluates this fuzzy look-up table instead of the rule set itself.

If the fuzzy controller only has few input variables, the utilization of a "fuzzy look-up table" is often the most suitable way. But if it possesses of a lot of input variables, the controller can not evaluate the table fast enough as it is too big. Perhaps the fuzzy look-up table does not approximate the control surface sufficiently. An additional drawback of a "fuzzy look-up table": It is not possible to adjust and validate a fuzzy controller evaluating a look-up table "on-line" at the system to be controlled - when having changed any rule or any membership function the fuzzy look-up table has to be updated again.

Therefore one often is restricted to the first approach. One has to solve the task to store the linguistic rule set in the available hardware environment. This rule set may consist of numerous rules as being demonstrated by the following example: Let the fuzzy controller have three input variables and let five linguistic values be defined for each input variable. A complete rule set may have up to $5^3 = 125$ rules. "Complete" means that for each possible combination of linguistic values of different input variables one has defined a linguistic rule having this combination or a subset of it as its premise.

A large-sized rule set can not be stored "any old how" but only sophisticated - particularly if only few storage space is available. This may be the case if one is restricted to a small and cheap stored-programable control unit. An additional requirement is that the controller must evaluate the rule set sufficiently fast.

2 A Way to Store a Linguistic Rule Set Efficiently

Let us consider a fuzzy controller with n input variables $X_1,...,X_n$ and one output variable Y. The limitation to one output variable is not a real restriction: One can replace a fuzzy

controller with m output variables by m controllers each having one output variable. We assume that the premise of each linguistic rule is an AND-OR-combination of linguistic values of different input variables. An example of such a premise is:
X_1 is 'very big' OR [X_1 ist 'big' AND X_2 ist 'small'].
Each conclusion is a single linguistic value of the output variable Y. A rule which has a conclusion consisting of several linguistic values of Y is replaced by several rules each having a single value as its conclusion. If this procedure fails, one defines an additional linguistic value of Y replacing the combination of linguistic values of Y.

The first basic idea is to integrate all rules having identical conclusions into one single rule. The premise of this rule joining different rules is transformed to a "standard structure": After this transformation the premise has the structure
ST_1 OR ST_2 OR ... OR ST_k,
where for r = 1,...,k the statement ST_r

- o is either a single linguistic value of an input variable
- o or an AND-connection of linguistic values of pairwise different input variables.

If necessary a distributive law is applied. An example: The premise
X_1 is 'very big' AND [X_2 is 'big' OR X_2 is 'medium']
is equivalent to the standard structure premise
[X_1 is 'very big' AND X_2 is 'big'] OR [X_1 is 'very big' AND X_2 is 'medium'].

This transformation is not possible if a rule is defined which has a premise with an AND-connection of different linguistic values of the same linguistic variable. An example is the premise "X_1 is 'very big' AND X_1 is 'small' ". Therefore we assume that the engineer avoids such premises. To formulate such a premise does not appear to be useful anyhow.

Let $A^{(1)}$, ... , $A^{(L)}$ be the linguistic values of Y. After having transformed the original rule set one obtains a set of L linguistic rules. Let $k^{(i)}$ denote the number of OR-connections in the i-th rule premise. For i = 1,...,L the i-th rule has the standard structure:

IF [ST_1 OR ST_2 OR ... OR $ST_{k(i)}$] THEN [$Y = A^{(i)}$].
In order to simplify the following description we assume that $k^{(i)} >= 1$ holds, i. e. each linguistic value of Y occurs at least once in a rule's conclusion. Of course we allow the case $k^{(i)} = 1$ which means that no OR-connection occurs in the i-th rule premise but only one single linguistic value or one AND-connection does.

The second basic idea is to keep the rule premises and the rule conclusions seperate when storing the linguistic rule set. The linguistic values of Y are numbered, for example from 1 to L. With the help of these numbers the premise of a rule can be associated with its conclusion. The conclusion is characterized by a membership function for a linguistic value of Y; this membership function has to be stored. But how to store a rule premise having the standard structure described above?

The linguistic values of each input variable are coded seperately. Let n_j denote the number of linguistic values of X_j. Practically one numbers the values of X_j from 1 to n_j.

Let ST_1 OR ST_2 OR ... OR ST_k be a linguistic rule premise which is to be stored. For r=1,...,k ST_r is either a single linguistic value or an AND-combination of linguistic values. The number of input variables is n. ST_r is stored as an n-dimensional vector $(p_{r1},...,p_{rn})$ of nonnegative integers: $p_{rj} = s$ means that the s-th linguistic value of X_j occurs in ST_r. As we restrict ourselves to rule premises having the standard structure, ST_r possesses one linguistic value of X_j at the most. If no linguistic value of X_j occurs in ST_r, $p_{rj} = 0$ is assigned. By performing this coding method for r=1,...,k, one obtains k vectors each having n elements. They are integrated into a k x n - matrix (k rows, n columns). The r-th row represents the combination ST_r and the j-th column is assigned to the j-th input variable X_j.

As the output variable Y has L linguistic values, this procedure yields L matrices $P^{(1)}$, ... , $P^{(L)}$ which are denoted "premise matrices" in the following.

The following simple example illustrates this coding method. Let the fuzzy controller possess two input variables X_1 and X_2.

Three linguistic values are defined for X_1 as well as for X_2. The linguistic values of X_1 are: "X_1 is 'small' ", "X_1 is 'medium' ", and "X_1 is 'big' ". Those of X_2 are: "X_2 is 'small' ", "X_2 is 'medium' ", and "X_2 is 'big' ". These linguistic values are numbered from 1 to 3 in this order.

The first linguistic value of Y is "Y is 'very small' ". This value occurs in the conclusions of the following rules of the original rule set:

IF "X_1 is 'big' " THEN "Y is 'very small' "
IF "X_2 is 'small' " THEN "Y is 'very small' "
IF "X_1 is 'small' " AND "X_2 is 'medium' " THEN "Y is 'very small' ".

Trnsforming the rules into the standard structure yields this rule for the first linguistic value of Y:

IF [ST_1 OR ST_2 OR ST_3] THEN "Y is 'very small' ",
where ST_1 = "X_1 is 'big' ", ST_2 = "X_2 is 'small' ", and ST_3 = "X_1 is 'small' AND X_2 is 'medium' ".

The premise matrix $P^{(1)}$, which contains the premise of this rule, is a 3 x 2 - matrix with the elements

$$P^{(1)} = \begin{matrix} 3 & 0 \\ 0 & 1 \\ 1 & 2 \end{matrix}$$

The complete linguistic rule set consists of L rules. The premises of these rules are stored with the help of L premise matrices $P^{(1)}, \ldots, P^{(L)}$. For i=1,...,L let $k^{(i)}$ denote the number of OR-connections in the i-th rule premise. Therefore $P^{(i)}$ is a $k^{(i)} \times n$ - matrix. After the engineer has defined the linguistic rule set, the matrices $P^{(1)}, \ldots, P^{(L)}$ are calculated and can be utilized by the fuzzy controller at run-time.

In the following we determine exactly the space required for storing a linguistic rule set in the manner just described:

- o The membership functions for the linguistic values of the input variables are to be stored. This takes the space for $n_1 + \ldots + n_n$ membership functions.
- o L membership functions for the L linguistic values of the output variable Y must be stored.
- o The L matrices for the linguistic rule premises require

- the space for $\sum_{i=1}^{L} k^{(i)} * n$ nonnegative integers
- and L times the space for storing the structure, address, and size of a matrix.

In practical applications the engineer typically only defines few linguistic values of an input or output variable. It seems not to be a real restriction to demand that he/she only defines 7 different linguistic values for each input variable at the most. If this limitation is followed, each element of a premise matrix is an integer of the set {0,1,...,7}. Storing this element takes exactly 1 byte. Therefore it requires $\sum_{i=1}^{L} k^{(i)} * n$ bytes to store all premise matrices.

How to store efficiently membership functions? In the most practical applications one restricts oneself to trapezoidal membership functions including indicator functions, triangular, and shoulder functions. Every trapezoidal membership function can be stored memory-saving with the help of four real numbers $x_1 <= x_2 <= x_3 <= x_4$ determined such that $\mu(x_1) = \mu(x_4) = 0$ and $\mu(x_2) = \mu(x_3) = 1$ holds.

3 Run-Time Evaluation of the Linguistic Rule Set

For each time t at run-time the fuzzy controller has to determine the real number y(t), the output value for t. The value y(t) depends on $x_1(t)$, ... , $x_n(t)$, the input values for time t. In order to calculate y(t), the fuzzy controller interpretes the linguistic rule set. When utilizing the original fuzzy control mode, a fuzzy controller performs this task in the following way:

(1) For j = 1,...,n let n_j be the number of linguistic values of the input variable X_j.
Let $\mu_1^{(j)}, \dots, \mu_{n_j}^{(j)}$ be the membership functions for these values. Let T be the t-norm selected for calculating AND-

connections and let U be the t-conorm for OR-connections. Let N denote the number of linguistic rules.
The fuzzy controller calculates the numbers $\omega_1(t), \dots, \omega_N(t)$ where for i=1,...,N $\omega_i(t)$ is the truth value of the premise of the i-th rule. $\omega_i(t)$ depends on the premise, on the membership degrees $\mu_{1_j}^{(j)}[x_j(t)], \dots, \mu_{n_j}^{(j)}[x_j(t)]$ ($j \in \{1,\dots,n\}$), and of the choice of T and U.

(2) Let $\nu^{(1)}, \dots, \nu^{(L)}$ be the membership functions for the linguistic values of Y. The selected fuzzy inference mechanism is characterized by a function F : [0,1] X [0,1] → [0,1]. Besides other functions the usual minimum function, the algebraic product, and Gödel's implication are utilized by fuzzy controller.
For i = 1,...,N let $\omega_i(t)$ be the truth value of the i-th rule premise. Let $\nu^{(s_i)}$ be the membership function for the conclusion of the i-th rule where $s_i \in \{1,\dots,L\}$ for $i \in \{1,\dots,N\}$. Evaluating the i-th rule yields the membership function $\zeta_{i,t}$ with
$\zeta_{i,t}(y) = F[\,\omega_i(t), \nu^{(s_i)}(y)\,]$ for $y \in W_Y$ where W_Y is the domain of Y.
If one integrates rule with identical conclusions into one rule, N = L and $s_i = i$ für $i \in \{1,\dots,L\}$ holds.

(3) With the help of the t-conorm U the N membership functions $\zeta_{1,t}, \dots, \zeta_{N,t}$ were summarized to one single membership function ζ_t.

(4) The fuzzy controller calculates the output value y(t) by the "defuzzification" of ζ_t.

The "defuzzification" is that subtask of one calculation step which consumes the most run-time. In order to avoid run-time defuzzification, different authors have suggested a pre-defuzzification, cf. [1, 5, 6, 7], e. g.
The basic idea of pre-defuzzification is: The membership functions $\nu^{(1)}, \dots, \nu^{(L)}$ are already "defuzzified" after having created the fuzzy controller and not during run-time. By this one obtains L real numbers $A^{(1)}, \dots, A^{(L)}$. For i=1,...,N the i-th conclusion "$Y = \nu^{(s_i)}$" is replaced by the crisp conclusion "$Y = A^{(s_i)}$". At run-time the fuzzy controller determines its output value by utilizing the formula

$$y(t) = \sum_{i=1}^{N} \omega_i(t) * A^{(S_i)} \; / \; \sum_{i=1}^{N} \omega_i(t) .$$

In [5] a combination of plausible conditions is presented which guarantees that $\sum_{i=1}^{N} \omega_i(t)$ always equals 1.

Sometimes the alternative formula

$$y(t) = \sum_{i=1}^{N} \omega_i(t) * \mathrm{area}^{(i)} * A^{(i)} \; / \; \sum_{i=1}^{N} \omega_i(t) * \mathrm{area}^{(i)}$$

is applied where $\mathrm{area}^{(i)}$ is the area of the membership function $v^{(i)}$.

In chapter 2 a mechanism for storing a linguistic rule set is described. This mechanism can be utilized for both basic inference modes. After being transformed into the standard structure, N = L holds. The fuzzy controller performs the work of step (1), i. e. the calculation of $\omega_1(t), \ldots, \omega_L(t)$, in three partial steps:

(1.1) It "fuzzifies" the n input variables, i. e. it calculates for j=1,...,n the present membership degrees of $x_j(t)$ to all membership functions representing linguistic values of X_j. By this the fuzzy controller calculates the vector $\{ \mu_1^{(j)}[x_j(t)], \ldots, \mu_{n_j}^{(j)}[x_j(t)] \}$.

(1.2) It arranges the membership degrees according to the rule premises: At run-time it calculates for i = 1,...,L the $k^{(i)} \times n$ - matrix $A^{(i)}(t)$ applying the formula:
$\alpha_{rj}^{(i)}(t) = 1$, if $p_{rj}^{(i)} = 0$ and
$\alpha_{rj}^{(i)}(t) = \mu_s^{(j)}(t)$, if $s := p_{rj}^{(i)} > 0$.
$\alpha_{rj}^{(i)}(t)$ and $p_{rj}^{(i)}$ are the integer-valued element of the r-th row and the j-th column of $A^{(i)}(t)$ and $P^{(i)}$, resp. Whereas the controller calculates $A^{(i)}(t)$ for each sample point again, the premise matrix $P^{(i)}$ has already been created before the run-time and is not changed at run-time.

(1.3) For i = 1,...,L the controller calculates the truth value $\omega_i(t)$ of the i-th rule premise according to
$\omega_i(t) = U [S_1^{(i)}(t), \ldots, S_{k(i)}^{(i)}(t)]$,
where for $r = 1,\ldots,k^{(i)}$

$s_r^{(i)}(t) = T\,[\,\alpha_{r1}^{(i)}(t), \ldots, \alpha_{rn}^{(i)}(t)\,]$
is the truth value of the r-th partial statement in the premise of the i-th rule.
In the first step the fuzzy controller applies the t-norm T to each row of $A^{(i)}(t)$. In the second step it applies the t-conorm U to all $k^{(i)}$ numbers obtained by the first step. If one selects the t-norm "algebraic product" and the t-conorm "algebraic sum", it holds:

$$\omega_i(t) = \sum_{r=1}^{k^{(i)}} \prod_{j=1}^{n} \alpha_{rj}^{(i)}(t).$$

Fig. 1 illustrates the work of the fuzzy controller.

Let us consider the simple example of the previous chapter. Let $x_1(t)$ and $x_2(t)$ be the present input values. The partial step (1.2) yields:

$$A(1) = \begin{matrix} \mu_3^{(1)}[x_1(t)] & 1 \\ 1 & \mu_1^{(2)}[x_2(t)] \\ \mu_1^{(1)}[x_1(t)] & \mu_2^{(2)}[x_2(t)] \end{matrix}$$

By partial step (1.3) the expected result is obtained:

$$\omega_1(t) = U\,\{\,\mu_3^{(1)}[x_1(t)],\ \mu_1^{(2)}[x_2(t)],\ T[\,\mu_1^{(1)}[x_1(t)],\ \mu_2^{(2)}[x_2(t)]\,]\,\}$$

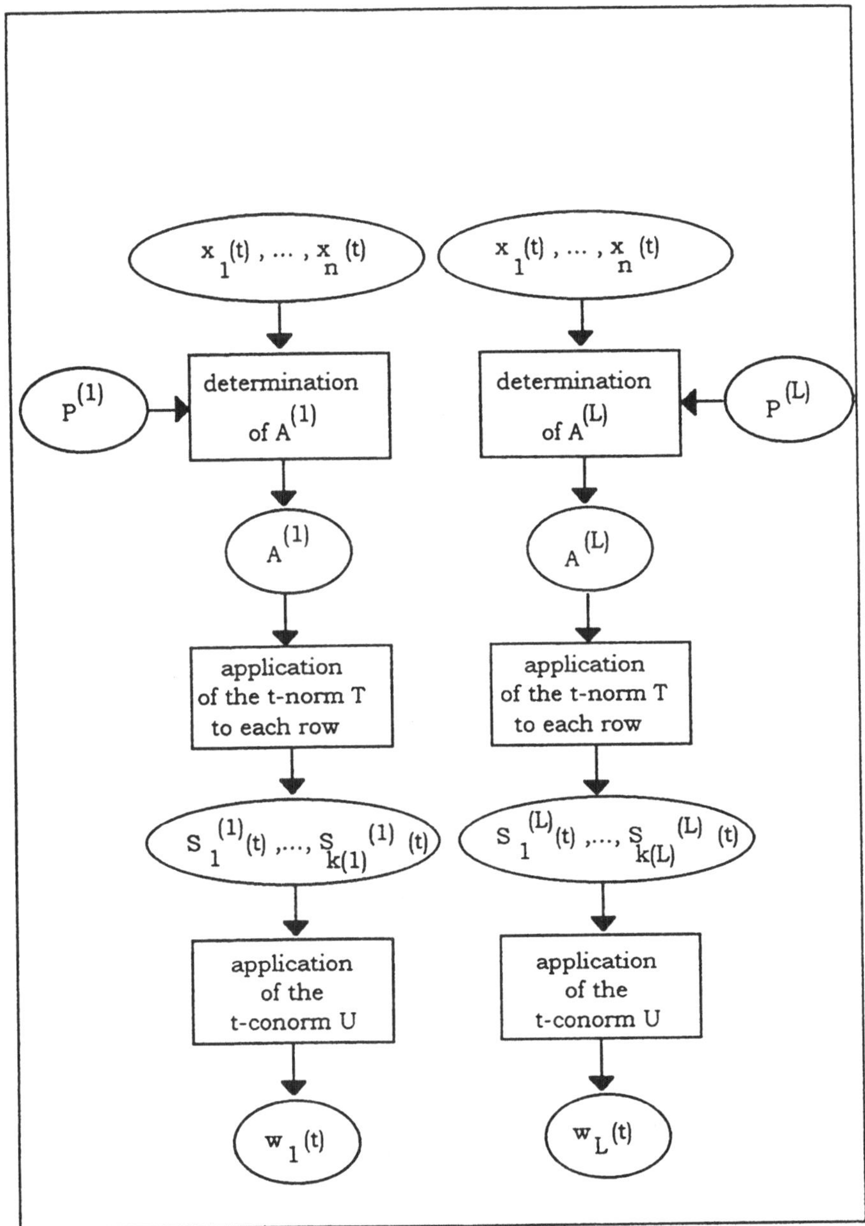

Fig. 1. Determination of the truth values $\omega_1(t), \ldots, \omega_L(t)$ by evaluation of the premise matrices $P^{(1)}, \ldots, P^{(L)}$

4 A Possible Application: Standard Functional Units

Often one plans automatization systems by configuring them with the help of standard functional units of an automatization language. For the automatization languages DOLOG AKF (from the AEG) as well as for Step 5 (from Siemens) such standard functional units for fuzzy control are now available [2, 3, 8]. It is intended to realize the approach of this article within the language DOLOG AKF .

Fig. 2 illustates how one can configure a fuzzy controller with the help of standard functional units and data units. These example controller has three input variables X_1, X_2, and X_3 and one output variable Y. A bold rectangle stands for a functional unit and a light-faced one for a data unit.

- o Each membership funtion has a trapezoidal shape. The membership functions for the linguistic values of one input or output variable are stored together in a data unit of type MF. Each membership function is represented by four real numbers. If the fuzzy controller has n input and 1 output variable, one needs n+1 data units of type MF.
- o A premise being coded as described in chapter 2 is stored in a data unit of type REG.
- o The standard functional unit FUZZIFY has two inputs: Via the input X it receives the present value $x_i(t)$ of an input variable X_i. The second input, MF, is connected with the data unit for the membership functions for the linguistic values of X_i. The output LV yields the present membership degrees of $x_i(t)$ to the membership functions as a vector.
- o As the linguistic rule set possesses L rules, the fuzzy controller has L standard functional units of type RULE. As the fuzzy controller of our example has three input variables, every standard functional unit has three inputs LV1, LV2, and LV3 for the membership degrees of $x_1(t)$, $x_2(t)$, and $x_3(t)$,

resp. The fourth input REG of the standard functional unit RULE is connected with a data unit for the premise of these rule. The output LV yields the present truth value $\omega_i(t)$ of this premise.

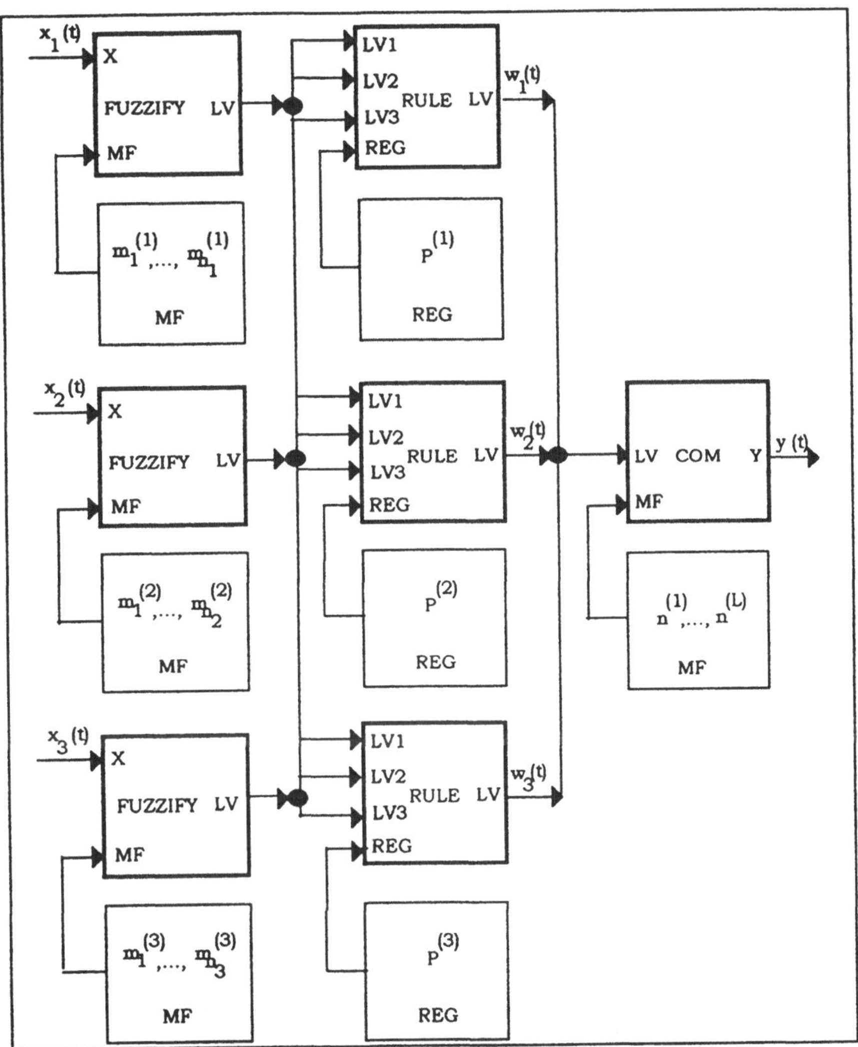

Fig. 2. An example for configuring a fuzzy controller with the help of standard functional units and data units. Bold rectangles: standard functional units

- o The present truth values $\omega_1(t)$, $\omega_2(t)$, and $\omega_3(t)$ are combined to a vector (LV[1], LV[2], LV[3]).
- o The standard funtional unit COM ("center of maximum") performs the "defuzzification". Via the input LV it obtains the present truth values of all premises, via the input MF the membership functions for the linguistic values of Y.

The solution just presented requires a specific standard functional unit for each number n being a possible number of input variables. This standard functional unit has n inputs LV1, ... , LVn. The standard functional unit of fig. 2 can only be used for controllers with three input variables.

If one wants to utilize the data units of type MF and those of type REG for all possible numbers of input variables, one should store the present number of input variables in the data units.

Acknowledgement:
The author thanks Mr. Achim Morkramer (AEG MODICON Europe, Seligenstadt, Germany) for several useful suggestions and hints.

References

[1] C. L. McCullough: "An anticipatory fuzzy logic controller utilizing neural net prediction". Simulation Vol. 58 No. 5 (1992), pp. 327 - 332.

[2] B. Cuno & K. D. Meyer-Gramann: "Anwender wollen keine Insellösungen - Fuzzy-Systeme eingebettet in Standard-Automatisierungskomponenten". Elektronik plus 2/1993, S. 40 - 44.

[3] B. Cuno, A. Morkramer & K. D. Meyer-Gramann: "Integration von Fuzzy Control in Geamatics, das Automatisierungssystem der AEG." Tagungsband VDE-Fachtagung "Technische Anwendungen von Fuzzy-Systemen" (1992), S. 101 - 110.

[4] H. KIENDL & J.-J. RÜGER: "Verfahren zum Entwurf und Stabilitätsnachweis von Regelungssystemen mit Fuzzy-Reglern". Automatisierungstechnik 41 (1993), S. 138 - 144.

[5] K. D. MEYER-GRAMANN & E.-W. JÜNGST: "Fuzzy Control - schnell und kostengünstig implementiert mit Standard-Hardware." Automatisierungstechnik 41 (1993), S. 166 - 172.

[6] M. MIZUMOTO: "Realization of PID controls by fuzzy control methods". Proceed. IEEE International Conference on Fuzzy Systems (1992), pp. 709 - 715.

[7] H.-P. PREUSS: "Fuzzy Control - heuristische Regelung mittels unscharfer Logik".
Teil I: Automatisierungstechnische Praxis 34 (1992), S. 176 - 184;
Teil II: Automatisierungstechnische Praxis 34 (1992), S. 239 - 246.

[8]: H.-P. PREUSS, E. LINZENKIRCHNER & S. ALENDER: "Fuzzy Control - werkzeugunterstützte Funktionsbaustein-Realisierung für Automatisierungsgeräte und Prozeßleitsysteme". Automatisierungstechnische Praxis 34 (1992), S. 451 - 460.

2.4. Learning the Rule Base of a Fuzzy Controller by a Genetic Algorithm

Jörn Hopf and Frank Klawonn

Abstract

For the design of a fuzzy controller it is necessary to choose, besides other parameters, suitable membership functions for the linguistic terms and to determine a rule base.
This paper deals with the problem of finding a good rule base — the basis of a fuzzy controller. Consulting experts still is the usual but time–consuming and therefore rather expensive method. Besides, after having designed the controller, one cannot be sure that the rule base will lead to near optimal control. This paper shows how to reduce significantly the period of development (and the costs) of fuzzy controllers with the help of genetic algorithms and, above all, how to engender a rule base which is very close to an optimum solution.
The example of the inverted pendulum is used to demonstrate how a genetic algorithm can be designed for an automatic construction of a rule base.
So this paper does *not* deal with the tuning of an existing fuzzy controller but with the genetic (re–)production of rules, even without the need for experts. Thus, a program is engendered, consisting of simple "*IF* ... *THEN* ..." instructions.

1 Introduction

In opposition to classical control techniques, which are mainly based on a mathematical model of the process to be controlled, the idea of fuzzy control is to model the behaviour of a (hypothetical) operator who is able to control the process. When the basic principles of the process and the reaction caused by the control actions are well understood, it is possible to design a reasonable control strategy using "*IF* ... *THEN* ..." rules. But in most cases it is necessary to consult an operator who is able to control the process and who has an intuitive understanding of the behaviour of the process. However, the knowledge acquisition process is often very difficult, since the operator is not always aware of all the rules he uses and might not be able to formulate an appropriate description of his control strategy. Therefore, a rule obtained from an operator might not work as well as expected. In some cases it is even impossible formulate a rule base, e.g. when a new process is to be controlled and no competent operator is available.

In this paper we propose the application of genetic algorithms to the problem of the design of a rule base for a fuzzy controller without the use of a priori information or the help of an operator.

The quality of the controller working with the rule base determined by the genetic algorithm depends on the choice of the evaluation function used in the genetic algorithm. In this way, we can prescribe the desired quality of the fuzzy controller to be generated.

The "IF ... THEN ..." rules appearing in the rule base of a fuzzy controller form an algorithm. Unlike a C–program, such an algorithm can be engendered by a genetic algorithm. The possibility of using an evolutionary process has its roots in the independence of these program modules. In genetic terms: changing the genotype does not affect the determination of the algorithm, what happens is only a phenotypical alteration – a changing of the algorithm's behaviour. The used language of a reduced instruction set code:

program	::	clause [clause ...]
clause	::	**IF** expression **THEN** expression
expression	::	variable **is** fuzzy–set [**and** expression]

Limited to:

clause	::	**IF** expression_1 **THEN** expression_2
expression_1	::	expression_2 **and** expression_2
expression_2	::	variable **is** fuzzy–set

2 The Control Problem

We consider the following control problem known as the inverted pendulum. A pole has to be balanced on a cart to be moved in the horizontal direction. A mass is fixed at the end of the pole. We neglect the frictional resistance here. The goal is to balance the pole by applying some force to the cart accelerating it. The force should be determined by the actual angle and angular velocity of the pole. The movement of the pendulum follows the differential equation

$$(m + M \cdot \sin^2\theta) \cdot l \cdot \ddot{\theta} + M \cdot l \cdot \sin\theta \cdot \cos\theta \cdot \dot{\theta}^2 - (M + m) \cdot g \cdot \sin\theta$$

$$= -F \cdot \cos\theta$$

where g is the gravitational constant, l the pole length, M the mass at the head of the pole, m the mass at the foot of the pole and $-90^\circ \leq \theta \leq 90^\circ$ is required.
The two parameters to be optimized are the angle θ and the angular velocity $\dot{\theta}$.

We use a Sugeno fuzzy controller (see f.e. [Kruse et al., 1994]) to solve this control problem. The fuzzy sets for the fuzzy controller are denoted as usual with Negative–Big (NB), Negative–Medium (NM), Zero (ZE), Positive–Medium (PM) and Positive–Big (PB). The membership functions of all fuzzy sets are chosen as triangular functions and are uniformly distributed over the universe of discourse. For the output of each rule we chose a fixed value which we also associated with a linguistic expression of the above mentioned type.

3 How a Genetic Algorithm Works

Genetic algorithms represent a strategy to search efficiently for near optimal solutions in difficult search spaces. Each solution is represented by one individual of a whole population. New solutions can be engendered by combining selected old solutions.

To solve a problem, a genetic algorithm requires five components: [Davis, 1987]

1. A representation of the solution to a problem in the form of a chromosome (chromosomal representation).
2. An initial population of individuals (solutions).
3. An evaluation function which indicates the *fitness* of each individual. This fitness shows how well the individual is able to cope with the given environmental factors.
4. Genetic operators which determine which genes of which parent will be passed on to their offsprings in the process of reproduction.
5. Parameters, as there are: the size of the population and probabilities employed by the genetic operators.

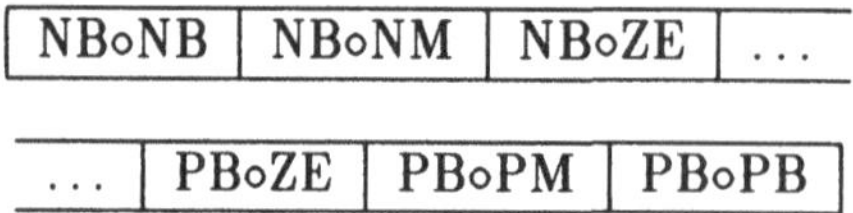

Figure 1: Chromosomal representation of the solution

A *population* corresponds in our application of genetic algorithms to fuzzy control to a family of rule bases. The *initial population* is chosen randomly.

The two most important genetic operators are *mutation* which means in our application of genetic algorithms to fuzzy control modifying a rule base by random, and *crossing over*, a recombination of two 'parent' rule bases. For a detailed discussion of these operators see f.e. [Beightler et al., 1979], [Davis, 1987], [Dewdney, 1986], [Goldberg, 1989], [Holland, 1992] and [Michalewicz, 1992].

3.1 Genetic Encoding of the Possible Solutions

To solve a problem with the help of a genetic algorithm it is first of all necessary to encode a general solution of the problem in a chromosomal representation. This representation has been chosen in correspondence with our 2–dimensional rule panel. Transferred to a 1–dimensional representation a rule base looks as shown in figure 1. Each box stands for one gene and is indexed with the premise of its corresponding rule in figure 1. The possible alleles (values) for each gene are the linguistic expressions for the output value in the rule base of the fuzzy controller. Note that we refrain from a binary representation here.

It is now the task of the genetic algorithm to fill out the rule base in figure 2 in a way that it contains the appropriate rules. The fuzzy controller on the basis of these rules must be able to hold the pole in the upright position both as quickly as possible and with the least possible deviation.

3.2 Generating the Initial Population

Many conventional optimization/search procedures use only one single starting point. Its position in the search space determines the next step which again leads us to another single point. This method, however, incorporates the following problem. When a local optimum has been found the global optimum might remain inaccessible, as the algorithm would have to give up the local optimum found before.

A genetic algorithm starts with a randomly generated initial population of possible solutions, i.e. a genetic algorithm relies on a set of starting points. In our case each chromosome represents a complete rule base for the fuzzy controller and determines the individual's position within the search space. Some may sit in a low valley (poor solutions to the problem), others may be found on high mountains (good solutions). Thus, to search efficiently, the starting population should be spread evenly over the entire space.

First – and according to expectation – none of the individuals with its chromosomes will be able to solve the problem (holding the inverted pendulum upright) sufficiently. Now, before selecting and reproducing these individuals, the fitness of each individual must be determined by an appropriate evaluation function.

3.3 Evaluation

The evaluation function indicates the fitness of each individual (rule base) of the population and must be designed in such a way that the fittest individuals take part in the process of reproduction.

Along with an appropriate coding the evaluation function decides on the success of the genetic algorithm, see [Beightler et al., 1979], [Davis, 1987], [Dewdney, 1986], [Goldberg, 1989], [Holland, 1992] and [Michalewicz, 1992].

Each genetic algorithm requires a specific choice of the evaluation function taking the following aspects into account.

- For an evaluation function it is essential to assign high values to chromosomes that represent good solutions.

- On the other, an inhomogeneous gene pool with enough variety to find (near) optimal chromosomes (solutions) should be kept. Therefore, if the evaluation function favours good (but by far not optimal) chromosomes too strong against average chromosomes, the genetic algorithm will get stuck at some unsatisfactory solution.

Evaluation in this case is a tightrope walk between supporting those individuals with high fitness and keeping a rich variety in the gene pool.

(↓) angle \ (→) angular velocity					
force	NB	NM	ZE	PM	PB
NB	PM	NB	ZE	NM	PM
NM	NB	NM	NM	ZE	NM
ZE	NB	NM	ZE	PM	PB
PM	NM	ZE	PM	PM	PB
PB	PB	PM	PB	PB	PM

Figure 2: An engendered rule base

3.3.1 Evaluation Function for the Inverted Pendulum

In order to compute the value of the evaluation function for a chromosome for the problem of the inverted pendulum we run several times a simulation of the inverted pendulum controlled by the fuzzy controller on the basis of the rule base associated with the chromosome. For each simulation different values for the initial angle and the velocity are chosen. During this simulation score points for the evaluation can be gained by the chromosome taking the following criteria into account.

- A score can only be gained if the pendulum has been hold upright during the whole simulation.
- At the end of the simulation score points are granted for a small deviation from the upright position. The interpretation of small deviation is narrowed from generation to generation, since in the beginning usually no randomly generated chromosome will be able to hold the pendulum in the upright position.
- The time needed to reach a stable upright position of the pendulum is evaluated only indirectly by the chosen time of simulation.

The controller must be able to handle different initial situations. This is guaranteed by using a number of randomly chosen initial values for the simulation and evaluation. In an illustration, where good solutions are located on mountains and bad in valleys, this means an ever changing landscape. But although it is unlikely that the individuals are confronted with the same fitness landscape twice, an 'optimal' solution is engendered.

3.4 Genetic Operators

Genetic operators simulate changes of chromosomes in nature. Usually two operators are considered, namely cross over and mutation.

3.4.1 Cross Over

The cross over operator mixes the genes of two chromosomes in the phase of reproduction. In genetic algorithms cross over is realized in the following way. First of all pairs of chromosomes are selected from the population, usually randomly proportional to their fitness determined by the evaluation function. The idea of the cross over operator is to combine the features, especially the positive ones, of the chromosomes by mixing the genes of each pair of chromosomes. In this way new chromosomes are generated that replace their parents. Mixing of genes is achieved choosing one gene randomly — the cross over point — and exchanging the genes of the pair of chromosomes behind this cross over point. In order to illustrate the cross over operator we consider the (binary) chromosomes

1 0 1 0 1 1 0 0 1 0 1 0 1	0 1 1 1 0 0 1 0 0 0 1 0

and

1 0 1 1 0 0 1 0 1 0 0 1 1	0 1 0 1 0 0 1 0 1 0 1 0

The cross over point is chosen behind the 13th gene. Thus the cross over operator yields the following two chromosomes.

1 0 1 0 1 1 0 0 1 0 1 0 1	0 1 0 1 0 0 1 0 1 0 1 0

and

1 0 1 1 0 0 1 0 1 0 0 1 1	0 1 1 1 0 0 1 0 0 0 1 0

Since chromosomes to participate in cross over are chosen randomly, but with a probability proportional to their fitness, better chromosomes have higher chances to produce offspring by cross over.

As a side remark, we should mention that the cross over operator described above is the most simple one but also the one yielding the worst results [Michalewicz, 1992]. Therefore, we preferred in opposition to this one point cross over the two point cross operator where two genes are selected randomly and the gene sequences between these genes are exchanged.

3.4.2 Mutation

Besides cross over the other genetic operator is mutation. Whereas cross over mixes genes of different chromosomes and can in this way combine good solutions, mutation changes genes in one chromosome randomly. The reason for the use of the mutation operator is the following.

Mutation avoids the convergence to a population with a homogeneous gene pool and thus guarantees for a certain variety of genes. It should be emphasized that without mutation chances for a thorough search through the space of possible solutions are quite small. If a certain allele (value) for one of the genes is not present in the population, this allele cannot be generated by cross over. Therefore mutation is a necessary operator, even if it supports only a random search, not directly aimed to improve individual chromosomes.

Mühlenbein demonstrated [Mühlenbein and Schlierkamp-Voosen, 1993] that in most cases it is sufficient to work with a constant mutation rate (probability for changing one gene) of $R^M = \frac{1}{n}$, where n is the number of genes per chromosome.

3.4.3 The Building Block Hypothesis

The schema theorem (see f.e. [Michalewicz, 1992]) indicates that a genetic algorithm concentrates the search for an optimal solution on certain subspaces of the space of all possible solutions. These subspaces are characterized by schemata, chromosomes with undetermined genes. The subspace associated with a schema corresponds to the set of all genes that have the same alleles for those genes that are determined in the schema.

The building block hypothesis, derived from the schema theorem, states that a genetic algorithm mainly searches in those subspaces that are characterized by schemata with a short defining length, low order and high fitness. The defining length of a schema is the distance between the two outmost determined genes in a chromosome. The order of a schema is the number of determined genes. The fitness of a schema is the average fitness of all chromosomes in its corresponding search space.

For our problem – finding a suitable rule base for a fuzzy controller – it is easy to see that the building block hypothesis is applicable. If a rule base should be able to handle a certain situation, only a small subset of neighbouring rules is needed. Since the coding of the chromosomes maintains at least partly this neighbourhood relation, (partially) good rule bases correspond to schemata with a short defining length, low order and high fitness. Therefore, it is possible to generate an overall good rule base from two partially good rule bases by using the cross over operator.

4 Simulation Results

The best rule base (chromosome) obtained after 33 generations with a population size of 200 is illustrated in figure 2. Starting the simulation with an upright standing pendulum but with a high initial angular velocity the fuzzy controller using this rule base is able to balance the pole finally with a deviation of constantly less than one degree. The protocol of the simulation run is presented in figure 3, where the values for the angle, angular velocity, and the force are shown over the time. Note that this result is obtained by learning the rule base alone. The initial fuzzy sets remained unchanged.

The appendix shows a comparison between a fuzzy controller based on Łukasiewicz logic [Klawonn, 1992] designed by hand and the fuzzy controller obtained from the genetic algorithm.

Note that the genetic algorithm does only rely on the fitness of a rule base. Therefore, it is possible that the genetic algorithm finds a well working rule base that does not coincide with the rule base one would write down from an intuitive point of view. This is one of the reasons why the rule base in figure 2 is not symmetric. Some rules also nearly never apply and the entry in the table for these rule is not so important for the overall evaluation or fitness of the chromosome. This might also lead to deviations from the intuitively appealing rule base.

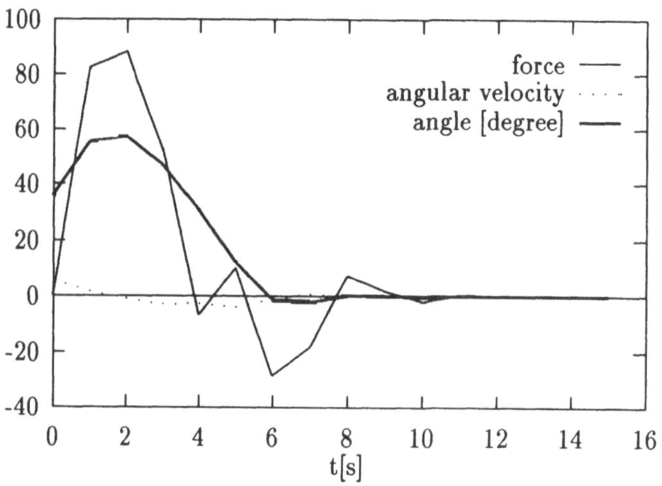

Figure 3: Change of angle, angular velocity and force due to a start impulse of 6.6 out of the neutral position.

References

C. S. Beightler, D. T. Phillips and D. J. Wilde (1979). *Foundations of Optimization*. Prentice-Hall, Englewood Cliffs, NJ, 2. edition.

L. Davis, ed. (1987). *Genetic Algorithms and Simulated Annealing*. Morgan Kaufmann, Los Altos, Ca.

A. K. Dewdney (1986). *Computer-Kurzweil*. Spektrum der Wissenschaft, pages 4–11.

D. E. Goldberg (1989). *Genetic Algorithms in Search, Optimization, and Machine Learning*. Addison–Wesley.

J. H. Holland (1992). *Genetische Algorithmen*. Spektrum der Wissenschaft, pages 44–51.

F. Klawonn (1992). *On a Łukasiewicz Logic Based Controller*. In *Proc. MEPP'92 International Seminar on Fuzzy Control through Neural Interpretations of Fuzzy Sets*, number 14 Ser. B. in *Reports on Computer Science & Mathematics*, pages 53–56, Turku, Finland. Åbo Academi.

R. Kruse, J. Gebhardt and F. Klawonn (1994). *Foundations of Fuzzy Systems*. Wiley, Chichester.

Z. Michalewicz (1992). *Genetic Algorithms + Data Structures = Evolution Programs*. Springer, Berlin.

H. Mühlenbein and D. Schlierkamp-Voosen (1993). *Optimal Interaction of Mutation and Crossover in the Breeder Genetic Algorithm*. Technical Report, GMD, Bonn, Germany.

R. Sommer (1992). *Entwurf und Implementierung eines auf Lukasiewicz-Logik basierenden Fuzzy Controllers*. Studienarbeit, Institut für Betriebssysteme und Rechnerverbund, Technische Universität Braunschweig.

Appendix

The following diagrams compare the action of a fuzzy controller on the basis of Lukasiewicz logic *(left)* [Sommer, 1992] with that of the controller engendered by a genetic algorithm *(right)*.
Initial values: angle = 40.0°, angular velocity = −2.0

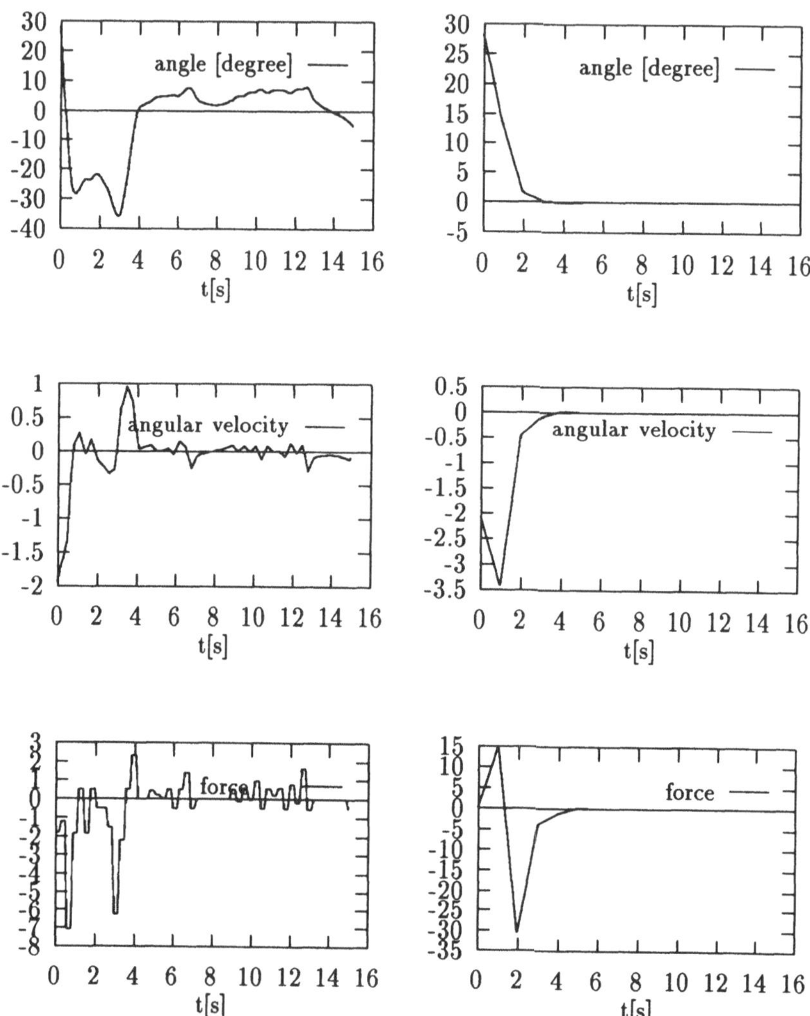

2.5. Comparison of Conventional and Fuzzy–Control of Non–Linear Systems

Heiko Knappe

Abstract

The results of conventional control theory are not really satisfactory in cases of non-linear systems of high order with uncertainties in parameter and structure. This is mainly due to the necessary reduction in order and simplification of the system. A fuzzy controller, on the other hand is not based on mathematical modells. Consequently, it is especially suitable for technical non-linear processes for which verbal control strategies are known. This paper presents the speed and position control of an elastic two-mass-system with slack. A fuzzy controller is compared to the conventional theory. A systematical fuzzy controller design as well as the robustness of the system concerning uncertainties in parameter and structure are researched.

1 Introduction

Usually it is not sufficient to merely concentrate on the engine when controlling the speed or position of an electrical drive. In reality a totally fix connection between drive and machine are not given (figure 1). In a first step the ideal situation of a fixed state is assumed. If the results are not satisfactory, the elastic shaft connection and possibly other non-linearities such as *slack* or *Couloumb friction* in transmissions, clutches and bearings have to be taken into concideration. Then a modell of an elastic two-mass-system has to be used. In many situations a fix shaft connection can be assumed. When dealing with the control of machine tools or robots the system should be modelled with two or even more masses. The more flexible the shaft and consequently the system is, the higher the influence of non-linearities, friction and slack.

The consequences are stick-slip limit cycles. Especially in multi-mass systems it is almost impossible or demands enormous effort to avoid these phenomena with conventional control theory. The position control of a two-mass system, for example, is based on a differential equation system of *6*th order when using an *I*-(integral) state controller. A complete mathematical model becomes impossible as soon as additional information, such as information about the production process (e.g. the request profile for the set values) is to be considered. In this case, fuzzy logic in combination with classical controller design constitues a new possibility for quicker design of better controllers. Figure 1 shows the model used for the two-mass system. A current- controlled dc-shunt-wound motor is used for the drive. For simulation purposes, the motor torque M_M can be used as control variable, since it shows a *PT2*-behaviour with a much smaller time constant than the mechanical system. Now a model for the two-mass system is designed. Based on this model state and fuzzy control concepts for speed and position control are developped and compared to each other. It is the goal of the project to find design methods for fuzzy controllers and compare the results to conventional techniques.

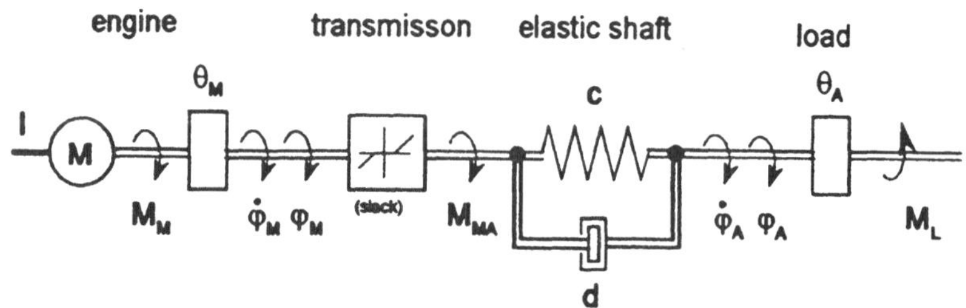

Figure 1: Elastic two-mass-system with slack
(model with an elastic connection of engine and machine in an electric drive system)

For the model of the system the non-linear slack is neglected in the beginning. The shaft, that connects drive and machine is assumed as mechanical torsion spring with the stiffness *c* and damping *d*. A system of third order is given. The basic equations for drive mass, shaft and machine mass lead to the differential equation system for the three state variables $\dot{\varphi}_M$, $\dot{\varphi}_A$ and $\Delta\varphi$. The engine torque M_M is the control variable and the load moment M_L constitutes the disturbance variable.

2 Design of State Controllers

The dynamic behaviour of the system is defined by the eigenvalues of the closed control loop. With state controllers one has the possibility to obtain a defined behaviour by defining suitable eigenvalues for the closed loop (e.g. quickly reaching the command variable with defined overshoot or asymptotic behaviour without overshoot). The two-mass-system allows an unlimited definition of eigenvalues, since it is totally controllable. For the simulation it is assumed, that all states of the system are measurable. Otherwise observers have to be introduced. For the design of a *state controller* it is necessary to have a complete *state description*:

$$\begin{bmatrix} \ddot{\varphi}_M \\ \Delta\dot{\varphi} \\ \ddot{\varphi}_A \end{bmatrix} = \begin{bmatrix} -\frac{d}{\theta_M} & -\frac{c}{\theta_M} & \frac{d}{\theta_M} \\ 1 & 0 & -1 \\ \frac{d}{\theta_A} & \frac{c}{\theta_A} & -\frac{d}{\theta_A} \end{bmatrix} \cdot \begin{bmatrix} \dot{\varphi}_M \\ \Delta\varphi \\ \dot{\varphi}_A \end{bmatrix} + \begin{bmatrix} \frac{1}{\theta_M} \\ 0 \\ 0 \end{bmatrix} \cdot M_M + \begin{bmatrix} 0 \\ 0 \\ -\frac{1}{\theta_A} \end{bmatrix} \cdot M_L \qquad [1]$$

2-mass system

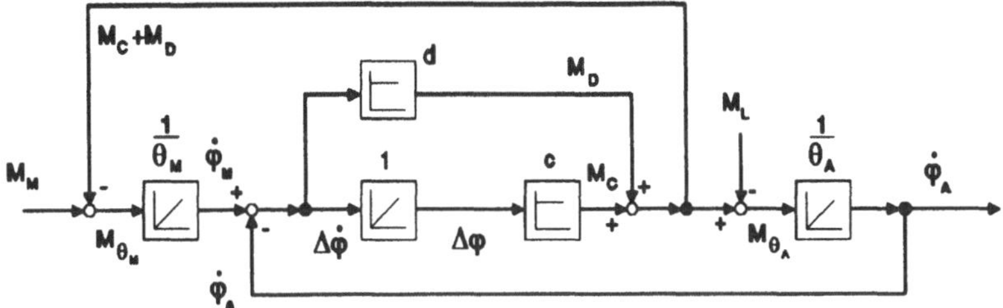

Figure 2: Non-normalized signal flow diagramm of the elastic two-mass-system

2.1 Speed Control

In a first step a *P*- (proportional) *state-controller* for the above system is to be defined. The states of the system n_1, n_2 and $\Delta\varphi_{12}$ are linearly fed back to the control variable with the coefficients k_1, k_2 and k_3. The closed control loop leads to a new equation system with a new system matrix $\mathbf{A}_Z$ that fully decribes the behaviour of the total system. Now the eigenvalues of the total system (of the pilot transfer function) are calculated in dependency of the *feed-back coefficients* k_1, k_2 and k_3. This happens with the help of the *characteristic polynominal*:

$$N_Z(s) = \det(s \cdot \mathbf{I} - \mathbf{A}_Z) = N_{Norm}(s) = s^3 + s^2 \cdot \xi_2 + s \cdot \xi_1 + \xi_0 \qquad [2]$$

The coefficients of the controller k_1, k_2 and k_3 are now defined by choosing a pilot polynominal and comparing the coefficients of the different polynominals with each other. The pilot polynominal defines the eigenvalues of the total control system. The desired behaviour of the controlled system is obtained by the choice of suitable eigenvalues. With the equation [2] the coefficients of the controller k_1, k_2 and k_3 can be expressed in dependency of the parameters ξ_1, ξ_2 and ξ_3. The prefiltering factor K_v is calculated from the condition of the stationary state. In order to obtain an asymptotic behavior, for example, a threefold real pole has to be used. Consequently the pilot polynominal for the asymtotic borderline case can be expressed as:

$$N(s) = (s + 1/T)^3 \qquad [3]$$

The solution of [2] leads to the equations for the coefficients k_1, k_2, k_3 and K_v in dependency of T. The time constant T can be choosen arbitrarely. The closer the threefold pole is to the point of origin or the larger T is defined, the faster the system becomes. The limitation for the size of T is, that as there is an increase of the necessary adjusting range, the mechanical (e.g. maximum permissible torsion angle $\Delta\varphi_{12}$) and electrical (e.g. maximum engine current) stress to the system also increases.
In drive theory, often the *damping optimum* (damping $D=0.5$) is used for controller adjustment. From the corresponding pilot polynominal the controller coefficients are obtained for the pole definition in the same way: Also, in this case, the above described conditions for the definition of the time constant T of the speed control loop have to be considered. Since the P-state controller does not have an integral part, it is not stationary exact for load moments (disturbances). Consequently an I-state controller was designed. It, however, is not able to follow an excitation as fast as the above described P-state controller. Figure 3 shows the results of the simulation of the speed controlled elastic two-mass system (eigenfrequency f_{12} $=100$ Hz) with the designed I-state controller for guidance and disturbance excitation.

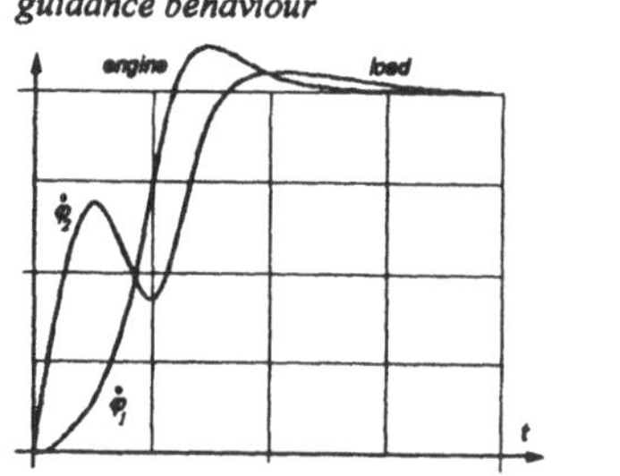

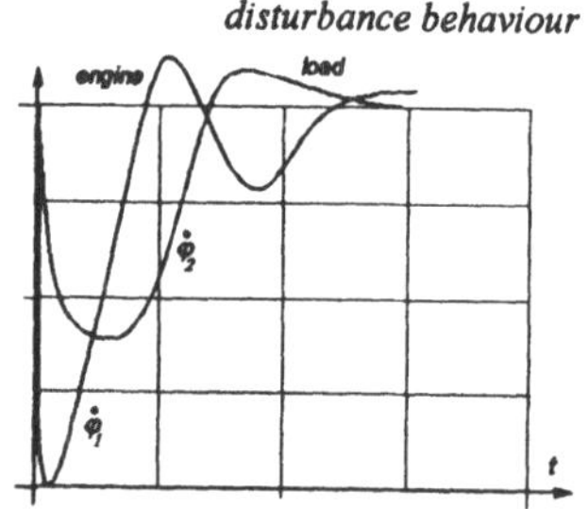

Figure 3: Results of the simulation of the speed control with an I-state controller

The time constant T of the state controller is chosen in such a way that the maximum permissible engine torque and the maximum permissible armature current are not exceeded. This makes it possible to have a direct comparison of the results of the state controllers with such of fuzzy controllers.

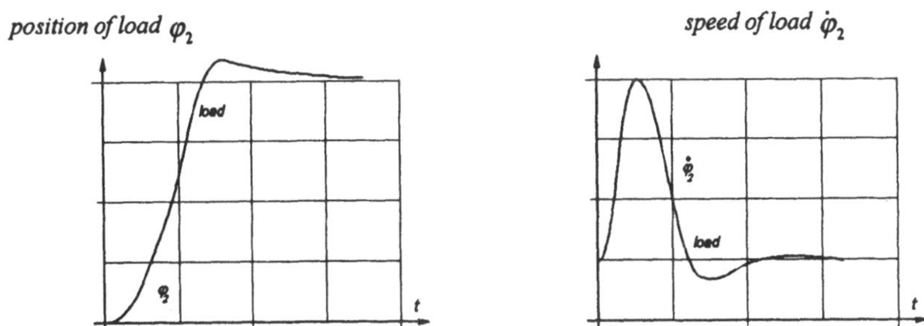

Figure 4: Results of the simulation of the position control with a P-state controller

2.3 Position Control

In order to be able to obtain the current angle φ_2 (= the position of the load mass) from the speed $\dot{\varphi}_2$, an integrator has to be added to the system in figure 2. Consequently, the system now becomes a system of forth order. An additional feed-back coefficient has to be introduced and calculated for the new state φ_2 for the P- or I-state controller. Figure 4 shows the result of the simulation for the position control with a P-state controller, that was optimized for the damping optimum.

2.4 Parameter- and Structure Uncertainty

The eigenfrequency f_{12} determines the behaviour of the system. The larger (smaller) the eigenfrequency, the harder (softer) is the system. This results form a lesser (greater) possible torsion of the shaft. This way, f_{12} also defines the speed of the controller. The softer the system, the lower is the possible controller speed in order to avoid extensive torsion of the shaft and the resulting oscillation of the two masses against each other.

So far, the system was simulated with $f_{12} = 100Hz$. Now the robustness of the I-state controller is to be scrutinized on the basis of speed control. For this purpose the controller that was optimized for the *100 Hz* system now is tested for the speed control of a two-mass system with a softer (*50 Hz*) and a harder (*150 Hz*) shaft. (figure 5). As expected, the transient behaviour response and the size of the overshoot improve and worsen for the harder (*150Hz*) and softer (*50 Hz*) system correspondingly. The oszillation of the masses is a sign that the controller is not optimized for the parameters of the system anymore. For the mechanically harder system the maximum permissible engine torque is a limiting factor. In the softer system, on the other hand, the maximum permissible torsion of the shaft has to be considered.

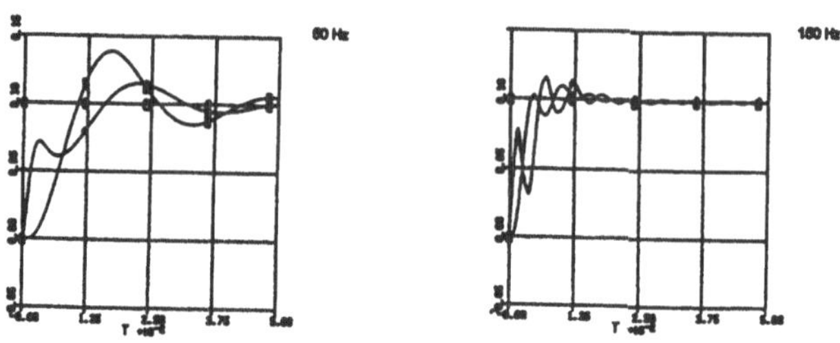

Figure 5: Simulation of the speed control with a P-state controller

The position control system is used for researching the effects of structure uncertainties. Therefore, slack is introduced additionally. The guidance behaviour of the P-state-position controller is shown in figure 6 for a step function response for a *5°* and *10°* slack.

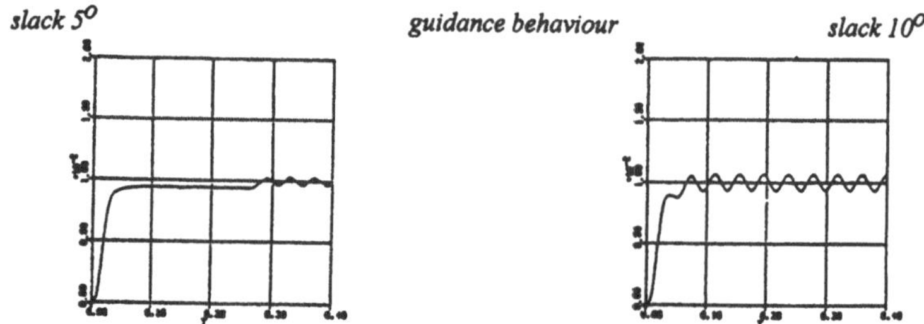

Figure 6: Simulation of position control with a P-state controller, considering slack

It has to be kept in mind that the controller was optimized for aperiodic behaviour and for a system neglecting slack. It can be observed, that the slack causes a limit cycle. The amplitude is dependent on the amount of slack. The frequency is defined by the eigenfrequency f_{12} of the system. The use of a state controller that was optimized for the damping optimum is even more critical. Through the typical overshoot the slack area is generally passed. This causes a continuous oscillation with a high amplitude. The results of the I-state controller are even worse. By integrating the diviation between the desired and the actual value, large regulation moments are created, resulting in some high amplitude oszillation. The sensitivity of the asymptotic P-state controller is mainly due to the large feed-back coefficient for the angle φ_2, that is derived from the calculation. Only a non-linear consideration of the slack e.g. through switching between different controllers (structure and/or parameter) can improve the behaviour of the controller.

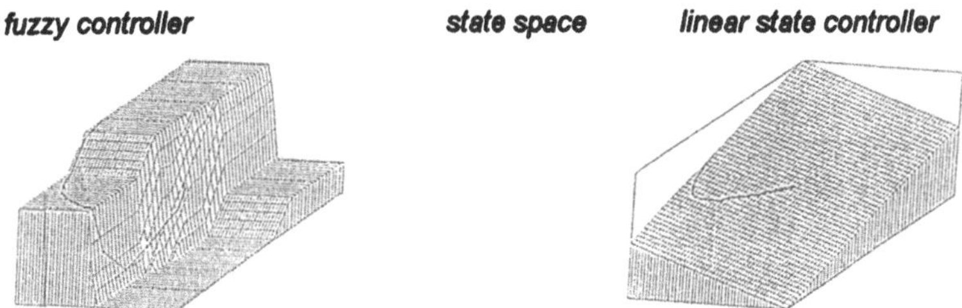

Figure 7 :State space of a linear P-state controller and a fuzzy controller

3 Fuzzy-Control

Finally, fuzzy-control systems have been defined for the speed and position control of the elastic two-mass system. In comparison to the traditional controller synthesis, fuzzy logic does not need any mathematical model of the system, but a description of the control strategy with verbal rules. Generally, a fuzzy controller is non-linear. Consequently a transient function response (phase-frequency characteristic) cannot be described like in linear systems. The fuzzy algorithm, therefore, constitutes a multi-dimensional non-linear transforming function between inputs and outputs. The advantage is that the controller can be adjusted optimally to the system. The disadvantage, however, is that no mathematical tools are available for optimizing. The design and optimization of a fuzzy controller, therefore, are only possible by experiment.

A fuzzy controller cannot be described through a closed function. It can, however, be shown in the *state space*. Figure 7 shows the state space of each, a linear and a fuzzy controller for two inputs and one output. In this case the controller is visualized through a three-dimensional state space. This state space is like a look-up table from which a certain output is defined for each input. Linear P-controllers always form a plain in the state space. In order to observe the behavior of a closed non-linear control loop for an excication function, the succeeding states (supporting points in a defined temporal order Δt) are connected to trajectories. The system dynamic can be followed from the density of the supporting points. The influence of the controller and the system to each other can be seen in the description of the state space. Consequently, it is a suitable aid for the optimization of the fuzzy controller. This method, however, is only possible for systems with merely two input varaibles. For systems of higher order a graphical description is not possible. If at all, cuts could be taken through the multidimensional description (two inputs are variable, all others are fix), but it does not make sense to create the trajectory. For a linear system merely the excitation function, but not its amplitude is of importance. In non-linear systems , on the other hand, this is different: The system reacts differently for step excications with different amplitudes. Caused by the fact that the trajectory changes with different excications , other locations of the phase state are passed through. As a consequence, different excications can cause totally different results. The system can, for example, show optimal behaviour for a step excitation. It, however, is not necessarily stable for a step-excitation with twice the amplitude. Since even a change of the system parameters cases different trajectories, this is also true for the robustness of non-linear controllers. Integral and differential components cannot be realized with a fuzzy system and have to be implemented outside the system. In this case, the description in the state space is even more difficult through an additional time dependency and additional states.

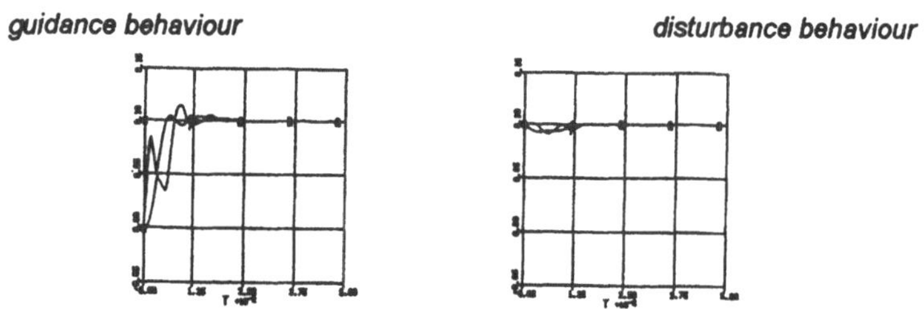

Figur 8: results of the simulation of the speed control with a fuzzy controller (PI-Fuzzy-Controller)

3.1 Speed Control

For the speed control of the elastic two-mass system fuzzy controllers were designed that consider all states of the system. Consequently, each rule contains three conditions and one conclusion. Figure 8 shows the results for guidance and disturbance behaviour. It was necessary to divide the linguistic variable into 7 subsets. In a first step a P-fuzzy controller was designed. (Pportion with non-linear characteristic ahead). As could be expected, it did not show stationary exactness during disturbance moments. A parallel linear integrator was introduced. The introduction of the fuzzy-I-portion led to a similar result as was received with the I-state controller.

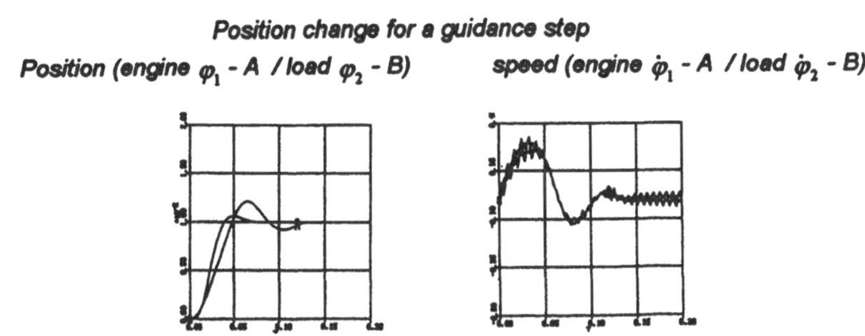

Figure 9: results of the simulation of the positon control with a fuzzy controllerr

3.2 Position Control

For the position control of the two mass system the development of a fuzzy controller that considers all *4* states was not performed due to the extreme effort. Instead, a position controller was superimposed over the existing optimized speed controller (cascade control). The position controller is based on the inputs of the rotational angle and speed of the load mass and calculates a desired speed for it. The inner speed-control-circle controls the engine torque from this desired speed. A worse result must be expected, since not all dependencies of the different states to each other can be considered. Moreover, the inner fuzzy controller is not optimized for random excitations as they are defined from the superimposed position controller. As a consequence, the simulation result in figure 9 shows a higher overshoot and a longer transient time in comparison to the conventional controller.

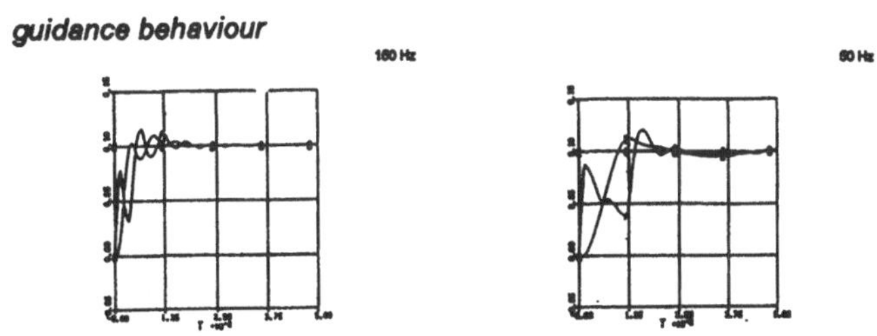

Figur 10: Results of the simulation of the speed control of the elastic two-mass system with the fuzzy controller (optimized for a system- eigenfrequency 100 Hz)

3.3 Parameter- and Structure Uncertainties

In order to test the robustness of the fuzzy speed control loop for parameter changes, the eingenfrequency of the system was changed. Figure 10 shows the results of the simulation for a system with *50 z* and *150 Hz*. As for the state controller it is also true: The softer the system, the worse the possibility to control it. The fuzzy controller shows a similar robustness as the linear state controller. For the test of the structure uncertainty, slack was now considered in the position-control-loop. Whereas the linear state controller already starts to oscillate (stabel oscillation) with a slack of *5°*, the fuzzy control loop is still stable with a slack of *10°*. (figure 11).

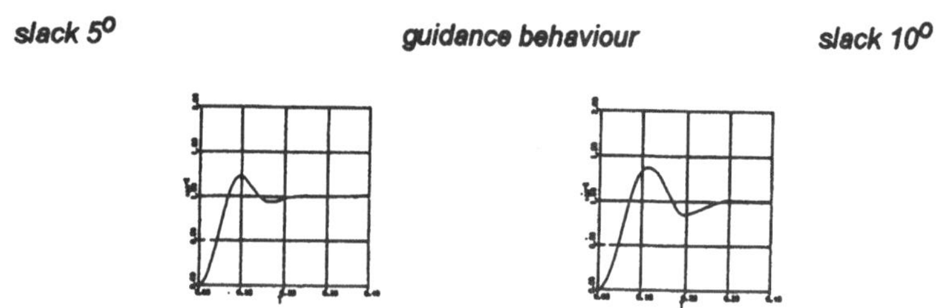

Figure 11:Results of the simulation of the fuzzy position control considering slack (load φ_2)

4 Summary

The state control proved to be quite robust for parameter changes. When slack was introduced, however, limit cycles were observed. In order to eliminate these, complicated structure and/or parameter switching strategies were necessary. The designed fuzzy controller on the contrary was robust for uncertainties in structure and parameters. This is due to the non-linear design, which gives filtering abilities to the fuzzy controller, that are adapted to the process behaviour in certain areas of the state space. The robustness is the result of a suitable choice of rules, subsets etc. It is, however, not true, that fuzzy systems generally are more robust. In the contrary, initial instable behaviour for changing excitations is usual. It is not possible to make general statements about the quality, robustness and stability of fuzzy systems, like this can be done for linear systems.

The fuzzy controllers were designed iteratively in sequencies of simulations and compared to linear controllers. Practical experience shows that it is helpful to reduce the amount of parameters for adjustment of a fuzzy controller. Only linear Sum-1-Subsets are used for the input variables to get a smooth control function. For the outputs singeltons are sufficient because

for defuzzification the maximum method *COM* was used. The minimum was used as *AND*-connection and the maximum as *OR* -connection. This is possible, since compensation is avoided already when creating the rule base for applications of control theory. These simplifications led quickly to an implementation of fast fuzzy controllers. Optimizing the controller, on the other hand, proved to be timeconsuming and difficult.
The creation of suitable rules is difficult for systems of high order (with several inputs). In this case, it is helpful to introduce intermediate values (cascade control systems). This, however, makes it impossible to consider the dependencies of all inputs. The controller can only work suboptimal. The design and optimization of the controller should be started at the inner control loop (e.g. speed control) and be performed step by step. Since we are dealing with a nonlinear system, the optimization is only possible for certain excitation functions (e.g. step functions). In case the superimposed fuzzy controller (e.g. position controller) determines the desired behaviour for the total system, usually there are random excitaiton functions. The danger arises that the inner control loop becomes suboptimal or even instable. In these cases a combination with linear controllers for the subimposed control loops is helpful.
The most positive factor of a fuzzy controller is, that no mathematical modell is needed. Traditional controllers usually are merely designed for one operating point. Consequently, complex control stategies can only be realized with great effort through switching between different controllers. With fuzzy logic, on the other hand, nonlinear systems can be controlled for more than one operation point. Even the most complex systems can be realized with the new theory. In addition, by using verbal fuzzy rules, also vague information can be considered. Moreover, due to their nonlinearity, fuzzy control systems (that are designed correctly) can show higher robustness than conventional controllers. Considering the fact, that mainly *PID* controllers are used in the industry up to now and the *PID* parameters are determined heuristically, everything suggests the implementaiton of the new fuzzy theory. Despite these advantages, one, however, should not forget the negative aspects: Even for the design of the structure of a fuzzy controller it is necessary to know the basics of control theory. It, for example, helps to know, that the I-portion (integrator) is responsible for the compensation of stationary diviations. Moreover, it is important to be familiar with methods of disturbance compensation and for damping the system. With the possibility of verbal controller design, there is the temptation to not even bother to develop a mathematical model, even for simple systems in which the use of a a linear controller would be absolutely

sufficient. Also, for traditional controller design heuristic design methods are necessary. This is especially true for nonlinear systems. Only an expert is able to choose from the variety of methods the optimal one for a specific process. The fuzzy solution, however, is a totally heuristic method that is based on experimenting. The adjustment of the membership functions for example, is done arbitrarely and by experiment. So far, no systematic methods are known for optimizing the controller, proving its stability and damping oscillations. It is not possible to dimension or estimate the parameters. Moreover, extensive (complete) tests are needed in order to verify the mode of operation (stability etc.) in a closed control loop.
Robustness and adjustments of the nonlinearities can only be obtained through thorough knowledge of the system. Consequently, also for the design of a fuzzy controller, some sort of model of the system is necessary. This model is not explicitly described in the form of a mathematical transfer function, but implicitly in the verbal description of the control strategy. The problem of the design of the rulebase should not be underestimated. Even if somebody is able to control a process surprisingly well without too much knowledge about it, this is mainly done subconsciously. Whereas the expression of *IF-THEN*-rules is very similar to human thinking, its development from subconscious strategies, however, is not easy.

Fuzzy-Control certainly is not a general solution that bursts all limitations of traditional control theory. The basic principles of physics cannot be turned around by fuzzy logic. Fuzzy is a new tool that allows the systematic realisation of rule-based strategies. This is, what fuzzy logic should be taken for. Whereas conventional methods will still dominate in control theory, fuzzy should be seen as a suitable and helpful tool for special problems. If control strategies are available in the form of verbal *IF-THEN* rules fuzzy logic should be used. Especially the combination with conventional controllers is very promising. In cases in which the description of the system and the number of inner states is unknown or not really measurable, the control of this system will always be only suboptimal. In reality, often *PID* controllers are used for such systems. The parameters of such a controller are determined subjectively by experiment. In cases of switching between different controllers the points of switching also have to be determined. Especially where PLCs (programmable logical controllers) are used, the approach to controller design is similar to the one of fuzzy control. In this case fuzzy logic is a powerful tool (e.g. mixture regulation [Knappe 93]) for the systematical implementation of rulebased control statements into an algorithm.

References

[Knappe 93 /1] H. Knappe, *"Fuzzy-Enteisungstechnik" ("Fuzzy De-/Anti-Icing Technique")*, Spektrum der Wissenschaft (Scientific American - German Edition), Page 97-98, 03/93 (1993)

[Knappe 93 /2] H. Knappe, *"Fuzzy-Controller for Flow and Mixture Regulation in an Aircraft Deicing Vehicle"*, IEEE Conference, San Francisco, Page 207-212 (1993)

[Knappe 92]. H. Knappe / Hrsg. B. Reusch, *"Regeln mit Fuzzy-Expertensystemen" ("Control with Fuzzy Expert Systems")*, 2. Dortmunder Fuzzy-Tagen, Page 131-142, Springer Verlag (1992)

[Lee 90] C. C. Lee, *"Fuzzy Logic in Control Systems: Fuzzy Logic Controller"*, Part I/II, IEEE Transactions on Systems, Man, and Cybernetics, Vol. 20, No. 2, March/April (1990)

[Sugeno 85] M. Sugeno, *"An Introductory Survey of Fuzzy Control"*, Information and Sciences 36, Page 59-83, Elsevier Publishing (1985)

[Takagi, Sugeno 85] T. Takagi, M. Sugeno, *"Fuzzy Identification of Systems and Its Applications to Modeling and Control"*, IEEE Transactions on Systems, Man, and Cybernetics, Vol. SMC-15, No. 1, Jan./Feb. (1985)

[Takagi, Hayashi 91] H. Takagi, I. Hayashi, *"NN-Driven Fuzzy Reasoning"*, International Journal of Approximate Reasoning, Page 191-212, Elsevier Publishing (1991)

[Zadeh 88] L. Zadeh, *"Fuzzy Logic"*, IEEE Computer, Page 83 - 93, April ´88 (1988)

[Zimmermann 91] H.-J.Zimmermann, *"Fuzzy Set Theory and its Applications"*, Kluwer Dordrecht, 2. Edition (1991)

3

Fuzzy Neuro Systems

3.1. Fuzzy Neuro Systems: An Overview

Detlef Nauck

Abstract

This paper gives an overview to different concepts of neural fuzzy systems. There are already several approaches to combine neural networks and fuzzy systems, to obtain adaptive systems that can use prior knowledge and that can be interpreted by means of linguistic rules as they are used e.g. in fuzzy controllers. Neural fuzzy models can be divided in two classes: Cooperative models which use neural nets and fuzzy systems separately, and hybrid models which create a new architecture using concepts from both worlds. Several of these approaches are discussed in this paper.

1 Introduction

The fascinating idea of combining neural networks and fuzzy systems is considered for several years, already. Ever since the application of the first fuzzy controllers the need for techniques to adapt a prototypical version of a controller to its task was recognized. But until today there are no sound theoretical foundations that can be applied to enhance the performance of an ill–defined fuzzy controller. Heuristic techniques that can be subsumed under the notion of *tuning* are used to change parameters in order to optimize its characteristics. Tuning can be a time–consuming process especially when the control task is complex.

Fuzzy controllers that are able to tune themselves by changing some of their parameters based on knowledge about the process were considered early. Approaches to so called *adaptive* or *self–organizing fuzzy controllers* can be found e.g. in [Procyk and Mamdani, 1979, Shao, 1988, Qiao et al., 1992]. An overview of this area is presented in [Driankov et al., 1993]. These kind of adaptive models usually use knowledge based methods and do not refer to neural networks.

Neural networks that are also called connectionist systems are designed to model certain aspects of the human brain. They consist of simple processing elements (neurons) that exchange signals along weighted connections. Certain types of neural networks are universal approximators, i.e. they can approximate any continuous function on a compact domain to any given degree of accuracy, and they accomplish this task by learning from examples. For an introduction to neural networks see [Aleksander and Morton, 1990, Nauck et al., 1994, Rojas, 1993]

The main advantage of neural networks, their ability to learn from examples, is reduced by their black box behaviour. Generally it is neither possible to use prior knowledge to initialize the network, nor their final state can be interpreted in terms of rules. Fuzzy systems on the other hand use knowledge, i.e. linguistic rules, and can always be interpreted in this way, but they are not able to learn. The advantages of one approach are mainly the disadvantages of the other one, and vice versa. So the idea comes naturally to combine neural networks and fuzzy systems to overcome their disadvantages, but to retain their advantages.

These combinations are called *neural fuzzy systems* or *neuro–fuzzy models*, and they form an area of research that enjoys constantly rising importance and interest in the "fuzzy systems community". Although most neural fuzzy systems are devoted to control issues, and they are considered as an important trend in fuzzy control (see the paper of H. Hellendoorn in this book), they can also be found in other domains, e.g. data analysis (see the paper of K.D. Meyer–Gramann in this book). There are also combinations that are concerned about optimizing the learning procedures of neural networks [Narazaki and Ralescu, 1991, Simpson, 1992a, Simpson, 1992b]. These models should be addressed as "fuzzy neural networks" or "fuzzy–neuro systems", but here they cannot be considered further. For an interesting approach of enhancing the performance of the backpropagation learning algorithm by fuzzy control see the article of Halgamuge, Mari, and Glesner in this book.

In this article approaches to neural fuzzy control systems are reviewed. There are cooperative models and hybrid models. Cooperative approaches use neural networks to determine certain parameters (e.g. fuzzy sets, or fuzzy rules) of a fuzzy controller which is then implemented without further using neural nets. Hybrid approaches define a new architecture that can be interpreted as a neural net and as a fuzzy controller. Their combination is therefore maintained during the whole lifecycle of the neuro–fuzzy controller.

2 Cooperative Neuro–Fuzzy Models

Cooperative neuro–fuzzy models can be usually characterized by one of four categories:

(a) **Learning fuzzy sets:** A neural net is used to learn fuzzy sets from sample data. This can be done by learning certain parameters characterizing a membership function, or by approximating the membership functions with neural modules which are then used within the fuzzy controller. Learning is usually done offline using a gradient descent algorithm. Fuzzy rules have to be developed apart from this.

(b) **Learning fuzzy rules:** A neural net develops fuzzy rules from sample data. This is usually done by offline clustering algorithms and the respective neural architectures are self–organizing feature maps, or similar approaches using winner–takes–all learning procedures, or adaptive vector quantization. Fuzzy sets usually have to be defined before the learning takes place. Instead of neural networks fuzzy clustering algorithms might be used [Bezdek and Pal, 1992].

(c) **Adapting fuzzy sets:** A neural net is used to change parameters of the membership functions of a predefined fuzzy controller. If there is an error measure describing the performance of the controller, this can be done online. Sometimes a neural network is not used explicetly, but a connectionist learning algorithm is applied directly to the fuzzy controller.

(d) **Scaling of fuzzy rules:** A neural net is used to determine weights for fuzzy rules. To do this fuzzy sets and fuzzy rules must be known. Learning can be done online or offline. The weights are usually interpreted as "importance" of rules [von Altrock et al., 1992, Kosko, 1992], and this implies some severe semantical problems (see Section 4). Scaling the output of a fuzzy rule is equivalent to change the membership function of its conclusion.

Next to these cooperative combinations there are approaches where a neural net is used as a pre–processor or a post–processor to a fuzzy controller. This is useful, if the input variables of the controller cannot be measured directly, and have to be created by a combination of numerous values. This way the neural net can function as an adaptive "information compressor". On the other hand the output of a fuzzy controller might not be directly applicable to a process, and has perhaps to be combined with other parameters. This combination can be realized by a neural network.

Within the context of this article these approaches are not considered as neural fuzzy models, because the neural net is not used to determine parameters of the fuzzy system.

An approach of type (a) is presented by Nokamura et al. Using supervised learning their model is able to determine the fuzzy sets of a Sugeno controller based on an existing fuzzy rule base [Nomura et al., 1992]. The antecedents of the rules use the parametrized triangular membership function:

$$\mu_r^{(i)}(\xi_i) = \begin{cases} 1 - \dfrac{2\,|\xi_i - a_r^{(i)}|}{b_r^{(i)}} & \text{if } a_r^{(i)} - \frac{b_r^{(i)}}{2} \leq \xi_i \leq a_r^{(i)} + \frac{b_r^{(i)}}{2} \\ 0 & \text{otherwise} \end{cases}$$

to represent the fuzzy set $\mu_r^{(i)}$ of variable X_i in fuzzy rule R_r. In this approach it is allowed to have fuzzy sets with $\mu_r^{(i)} \neq \mu_{r'}^{(i)}$, for $r \neq r'$, i.e. it is possible that the same linguistic term is represented differently in different rules. Because backpropagation is used to determine the parameters of the membership functions a differentiable t–norm is needed to evaluate the antecedents. Therefore Nokamura et al. use the product of the membership degrees, and not their minimum, to define the degree of fulfillment of a rule.

The problems of this approachs are the non–differentiable points in the triangular membership functions, and the potential semantical problems that arise by learning different representations for identical linguistic terms in different rules. For methods to overcome these disadvantages see e.g. [Bersini et al., 1993, Nauck et al., 1994].

The linguistic interpretation of self–organizing feature maps examined by Pedrycz and Card is a possibility to create fuzzy rules by connectionist learning [Pedrycz and Card, 1992], and it is a neuro–fuzzy model of type (b).

If the process that has to be controlled has n (input and output) variables the feature map consists of n input nodes, and the output layer is a two–dimensional map of $n_1 \times n_2$ nodes. For the competetive learning algorithm a set $\mathcal{L}$ of examples consisting of pairs of process state and correct control output values is needed. These values and the resulting weights are all from $[0, 1]$. After the self–organization of the feature map each process variable can be described by a single two–dimensional map that consists of the weight matrix $\mathbf{W}_i$ of the respective input node v_i.

After p_i fuzzy sets $\mu^i_{j_i}$ have been defined for each variable X_i, these fuzzy sets are used to transform the maps. This is done by selecting one fuzzy set for each variable. This selection is denoted as a linguistic description $\mathcal{B}$. The transformed maps are then intersected to obtain a matrix $\mathbf{D}^{(\mathcal{B})} = [d^{(\mathcal{B})}_{i_1,i_2}]$ that represents the compatibility of the learning result with the linguistic description $\mathcal{B}$ that represents the sequence $(j_1, \ldots, j_n)$:

$$\mathbf{D}^{(\mathcal{B})} = \bigcap_{i:i\in\{1,\ldots,n\}} \mu^{(i)}_{j_i}(\mathbf{W}_i), \quad d^{(\mathcal{B})}_{i_1,i_2} = \min_{i:i\in\{1,\ldots,n\}} (\mu^{(i)}_{j_i}(w_{i_1,i_2,i})).$$

$\mathbf{D}^{(\mathcal{B})}$ is a fuzzy relation, and $d^{(\mathcal{B})}_{i_1,i_2}$ is interpreted as the degree of support of $\mathcal{B}$ by node v_{i_1,i_2}. The height of $\mathbf{D}^{(\mathcal{B})}$ is interpreted as the degree of compatibility of $\mathcal{B}$ and the learning result. By describing $\mathbf{D}^{(\mathcal{B})}$ by its α–cuts $\mathbf{D}^{(\mathcal{B})}_\alpha$ one obtains subsets of output nodes whose membership degree is at least α, and by finding patterns $x_{k_0} \in \mathcal{L}$ for each v_{i_1,i_2} such that

$$||\mathbf{w}_{i_1,i_2} - \mathbf{x}_{k_0}|| = \min_{\mathbf{x}:\mathbf{x}\in\mathcal{L}} ||\mathbf{w}_{i_1,i_2} - \mathbf{x}||$$

holds, each $\mathbf{D}^{(\mathcal{B})}_\alpha$ induces pattern subsets $X^{(\mathcal{B})}_\alpha \subseteq \mathcal{L}$, and obviously

$$X^{(\mathcal{B})}_{\alpha_1} \subseteq X^{(\mathcal{B})}_{\alpha_2}, \text{ if } \alpha_1 \geq \alpha_2.$$

If there is a sufficiently large α_0 the induced set $X^{(\mathcal{B})}_{\alpha_0}$ can be interpreted as the set of prototypes of the class described by $\mathcal{B}$. Each $\mathcal{B}$ is a valid description of a cluster, if $\mathbf{D}^{(\mathcal{B})}$ has a non–empty α–cut $\mathbf{D}^{(\mathcal{B})}_\alpha$. Each $\mathcal{B}$ represents a linguistic control rule, and by examining each combination of linguistic values, a complete fuzzy rule base can be created.

This method also shows which patterns belong to no cluster, i.e. fuzzy rule. If their number is very high this may be due to an insufficient choice of membership functions. The problems of this approach lie in the determination of α_0 and the number units in the output layer. The learning algorithm is based on Kohonen's feature maps and its convergences is forced by reducing the learning rate, and there is no guarantee that the learning result properly represents the structure of the pattern set. Finally the learning result depends on the sequence in which patterns are propagated.

The advantages of this approach are that the structure of the pattern space is considered, and the information that is supplied by the membership functions is completely used to determine rules that fit the learning result best.

Another way to create a fuzzy rule base is presented by Kosko based on his FAM model (Fuzzy Associative Memory) [Kosko, 1992]. Assuming a finite domain $X = \{x_1, \ldots, x_m\}$ a fuzzy set $\mu : X \to [0,1]$ can be viewed as a point in the m-dimensional hypercube $I^m = [0,1]^m$. Then a fuzzy rule R: **If** ξ_1 **is** $A_{j,1}^{(1)} \wedge \ldots \wedge \xi_n$ **is** $A_{j,n}^{(n)}$ **Then** η **is** B_j can be interpreted as a mapping $R : I^{m_1} \times \ldots \times I^{m_n} \to I^s$.

A FAM is used to store a fuzzy rule (μ, ν) with a single antecedent μ and a single conclusion ν. Let $\mu_i = \mu(x_i)$ and $\nu_i = \nu(y_i)$, then a FAM is defined by its connection matrix

$$\mathbf{W} = [w_{i,j}] = \mu \circ \nu, \quad \min(\mu_i, \nu_j),$$

and it represents a fuzzy relation $\rho : X \times Y \to [0,1]$. $\mathbf{W}$ is denoted *fuzzy Hebb matrix* and their calculation is called *correlation minimum encoding* [Kosko, 1992]. The associative recall is defined as

$$\nu = \mu \circ \mathbf{W}, \quad \nu_j = \max_{i:i\in\{1,\ldots,m\}} \min\left(\mu_i, w_{i,j}\right) = \min(v_j, \text{height}(\mu)).$$

So the recall is always correct, if $\text{height}(\mu) \geq \text{height}(\nu)$ holds. This means there are no errors, if normal fuzzy sets are used, i.e. $(\exists\, x \in X)\ \mu(x) = 1$.

A FAM can only store a single fuzzy rule because a simultaneous storing of several rules would cause too many errors in the recall result. Due to the operations involved the disturbance between patterns would be even more severe then in common neural associative memories. To implement a rule with a conjunction within the antecedent, several FAMs have to be used, and their recalls have to be combined by minimum.

FAMs can be used to implement a fuzzy controller (see Fig. 1). The inputs to a FAM system consist of binary vectors where each of them has exactly one component with a value of 1 (fuzzy singletons). The output is a fuzzy set represented as a vector in $[0,1]^s$. If there is a defuzzification component involved, the output is also such a binary vector (BIOFAM: Binary Input-Output FAM). The encoded rules are additionally weighted by factors in $[0,1]$. Therefore a FAM system is a model of type (d).

The determination of the rule weights is done during the creation of the fuzzy rule base. The learning is done by a kind of adaptive vector quantization called *differential competitive learning* (DCL) [Kosko,

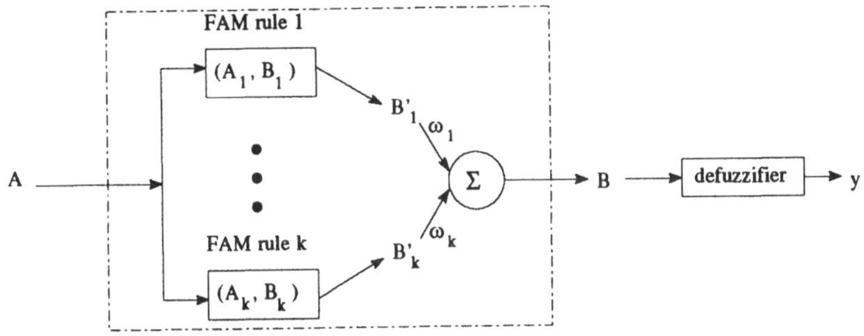

Figure 1: A FAM system [Kosko, 1992]: The A_i, B_i denote fuzzy sets, and the ω_i are weights from $[0,1]$

1992]. For this a neural network with n input units and $k = p_1 \cdot \ldots \cdot p_n$ output units is used. The input units are connected to all output units, and the output units are connected with each other to create a topology of lateral inhibition. This means one has to define a partioning on the variables, i.e. the fuzzy sets have to be known. The form of the membership functions has no influence on the learning result, because only their support is considered. Each output neuron , i.e. its weight vector, represents a possible fuzzy rule.

For the learning algorithm a set of examples of process states and correct control actions is needed. After the learning process for a fixed output unit and for all input units it is checked which fuzzy sets give a membership degree greater than zero for the weight between the respective input unit and the chosen output unit. This way a weight vector can be assigned to one or more rules. The weight of a fuzzy rule is determined from its number of weight vectors, e.g. by using the relative frequency. Now a FAM system can be created, and the learning process can be continued during the operation of the system by updating the rule weights, or by deleting or adding rules. An adaptive FAM system is a combination of the approaches of type (b) and (d).

The main disadvantage of this approach is the weighting of rules. First of all this produces semantical problems in the interpretation of a FAM system (see Section 4). If the set of examples used for the learning procedure is e.g. from observing a competent operator who controls the process under consideration, critical or extreme process state will be probably not sufficiently represented in the data set. If the learning goes on during operation the weights of these extreme states will become smaller and smaller resulting in a system that will be not able to control these states in the long run.

Kosko suggests to manually include rules to cope with extreme states, but this requires knowlegde about the process, and this might not be available. The user also has to make sure that the created rule base is consistent, i.e. that there are no rules with identical antecedents but different conclusions.

The FAM matrices cannot be learned, and they are a somewhat inefficient kind of storing fuzzy rules. The notion neuro-fuzzy system only fits to this approach because of the adaptive rule weights, and the rule learning capabilities. But the learning algorithm does not depend on the FAM representation, it can be used for any fuzzy controller. The learning algorithm also does not consider the topology embedded in the structure of the data set, as it is done by the model of Pedrycz and Card. The information encoded in the pre-defined fuzzy sets is also not fully used.

On the other hand the a FAM system is a simple and easy to implement model that offers some adaptive features. For this reason the approach is used in some commercial fuzzy shells. A similar approach that uses bidirectional associative memories (BAM) and the delta rule is used to control a flying helicopter model with four rotors [Yamaguchi et al., 1992]. This learning procedure directly changes the encoding of fuzzy rules in the assoziative memories and is therefore a model of type (c). For some commercial applications of cooperative neuro-fuzzy models in Japan refer to [Asakawa and Takagi, 1994].

3 Hybrid Neuro-Fuzzy Models

Hybrid neuro-fuzzy models create a new architecture by using concepts from neural networks and fuzzy systems. The models can usually interpreted in terms of a fuzzy systems, and they can be viewed as a neural network with special activation and propagation functions.

The ARIC model (Approximate Reasoning based Intelligent Control) by Berenji is a hybrid neuro-fuzzy model that uses several specialized neural networks (see Fig. 2). The architecture of ARIC uses concepts of adaptive critics, special neural controllers learning by reinforcement [White and Sofge, 1992], and it generalizes the neural model of Barto et

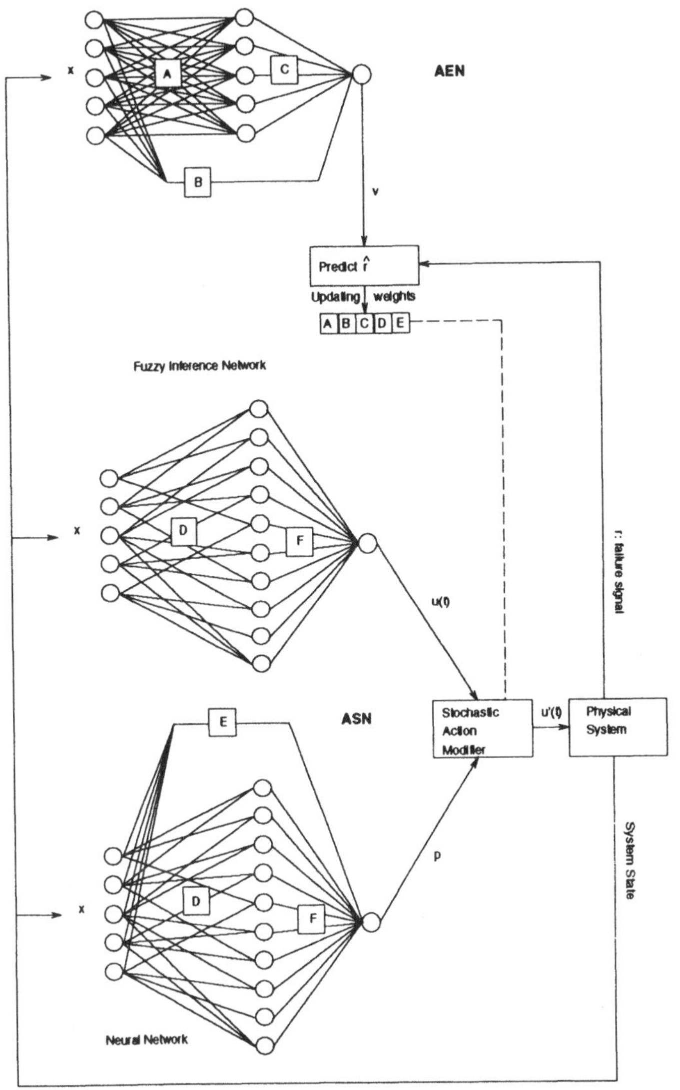

Figure 2: Architecture of ARIC [Berenji, 1992]

al. [Barto et al., 1983] to the domain of fuzzy control. ARIC consists of two neural modules, the ASN (Action Selection Network) and the AEN

(Action state Evaluation Network).

The ASN itself consists of two feedforward 3–layer neural networks. One of them is an direct representation of a fuzzy controller, and the other one calculates a confidence value that is used to change the output of the control network. The input layer of the control network represents the state variables of a process. Each unit stores the triangular membership functions of the respective variable, and they are connected to the units of the hidden layer which represent the fuzzy rules. The antecedent of a rule is represented by the linguistic values that are propagated on the weighted connection to the respective rule node, i.e. an input unit has to know which membership value has to be propagated on which connection. The membership values are scaled by the connection weights, and are combined within a rule node by the minimum function.

The membership functions of the conclusions are stored within the rule nodes. ARIC uses Tsukamoto's fuzzy sets, functions that are monotonous on their support. This way each rule delivers a crisp output value, because for each membership value greater zero there is exactly one element of the domain. The rule output values are propagated to the output unit that calculates a weighted sum using the weights of the connection between hidden layer and output layer. This output value is then scaled by a so called *stochastic action modification* based on the output values of the second (non–fuzzy) neural network of the ASN that uses the same weights as the control network with additional weights for connections between input and output layer.

The AEN network is an adaptive critic element that learns to predict the state of the process. Based on an external reinforcement signal that tells the AEN whether the process control has failed or not the network calculates an internal reinforcement signal that is used to adapt the weights in the whole ARIC system. If there is a high internal reinforcement (i.e. a good process state) the weights are changed such that their contribution to the output value is increased (rewarding). If the process control has failed the weights are changed such that their contribution is decreased (punishment). If the internal reinforcement is just small, the stochastic action modification is larger, allowing the system to randomly produce better ouput values. This approach is similar to adding Gaussian noise to the output values, as it is done in [Barto et al., 1983].

Because of the hidden layers used in the networks of ARIC the learning algorithm cannot be based on reinforcement learning alone. ARIC tries to optimize the internal reinforcement, and therefore concepts of backpropagation have been integrated into the learning procedure.

The disadvantages of ARIC are the complex neural architecture, and the semantical problems that arise by changing the weights in the control network. Changing the weights for the antecedents is equivalent to scaling the membership functions which results in non-normal fuzzy sets. Because it is not guaranteed that the weights are from $[0, 1]$ this may even result in functions that cannot be interpreted as membership functions anymore. Changing of the conclusion weights is equivalent to shift the membership functions which might produce undesired changes in the support of the fuzzy sets (e.g. Positive Big might drift to negative values). It is also possible that the same linguistic value is represented differently in different rules (see also [Nauck and Kruse, 1992, Nauck et al., 1993]).

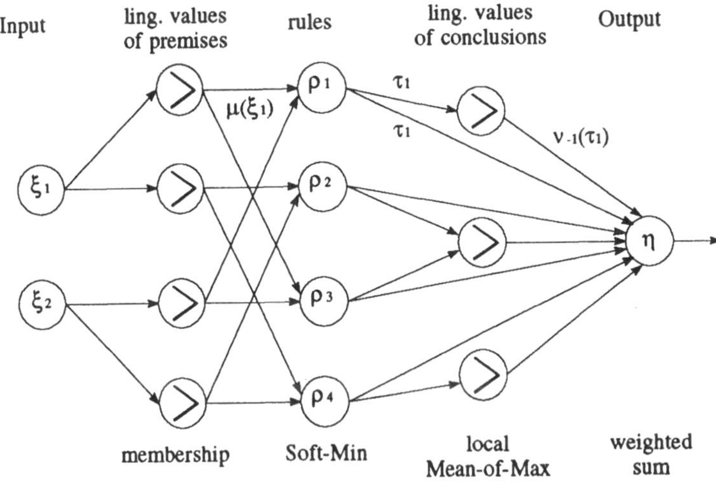

Figure 3: The ASN of GARIC [Berenji and Khedkar, 1992]

Berenji and Khedkar have tried to overcome the problems of ARIC by changing the ASN and defining the GARIC model (Generalized ARIC) (see Fig. 3). The ASN consists now of only one neural network that uses two additional layers of units to store the membership functions of the antecedents and the conclusions. The rule base is encoded by the connections, and there are no adaptive weights. The learning is done completely by adapting parameters of the triangular membership functions.

The learning algorithm of the AEN is unchanged, but the ASN learns by a kind of gradient descent to optimize the internal reinforcement signal. To do this a differentiable function to evaluate the antecedent of a rule is needed, i.e. the minimum function cannot be used here. GARIC uses a so called *soft minimum* function instead that is not a t-norm. Like ARIC the learning algorithm needs a crisp output value from each rule, i.e. it is not possible to use a defuzzification procedure on an aggregated fuzzy set determined e.g. by the usual maximum function. In GARIC a so called *local mean of maximum* procedure (LMOM) is used to obtain a crisp value from each rule, which only yields a result different from the usual MOM, if the membership functions are not symmetrical.

The learning algorithm uses gradient descent to optimize the internal reinforcement signal. But because the dependency of this signal from the control output calculated by GARIC is not explicetly known, the learning procedure has to use some heuristical assumptions. Additional problems that have to be heuristically solved are due to the three non-differentiable points of each membership function.

The learning algorithm depends on the changes in the internal reinforcement signal. If it is constant the learning stops. This situation occurs when the process is controlled optimally, but it might also occur when the process maintains a constant but non-optimal state. Therefore GARIC learns to avoid failure, and not to reach an optimal state. This may lead to an undesirable control strategy, because states short to control failure would be admissible. This kind of problem is addressed in [Nowé and Vepa, 1993].

Both ARIC and GARIC are just able to learn membership functions. The rule base of the controller has to be defined by other means. The models also need an initial definition of the fuzzy sets, and their number cannot change. This restriction usually holds for all hybrid neuro-fuzzy models. The advantage of these approaches is that no control values must be known for given states. The models learn by trial and error. This implies, of course, that a simulation of the process is available, or that learning can be done online at the process, what further implies that process failure is harmless.

Another approach of this kind is the ANFIS model by Jang (Adaptive Network based Fuzzy Inference System) [Jang, 1991, Jang, 1992]. This model is described by Brahim and Zell later in this book.

A model that completely consists of neural nets is the NNDFR model by Takagi and Hayashi (Neural Network Driven Fuzzy Reasoning) [Takagi and Hayashi, 1991]. This model uses neural nets to implement rules created by an external clustering procedure, and a neural net to determine the degree of fulfillment for each rule. The networks are trained by backpropagation. The model is not interpretable, i.e. there are no fuzzy sets implemented directly.

An approach that uses a similar architecture to the ASN part of GARIC is presented in [Sulzberger et al., 1993]. This so called FUN model (FUzzy Net) uses a stochastic learning procedure that randomly changes connections and parameters of membership functions, thus allowing the model to learn fuzzy sets and fuzzy rules. Because learning is done by a stochastic search process, and not by a connectionist learning algorithm, it is only marginally a neuro-fuzzy model.

The NEFCON model (NEural Fuzzy CONtroller) is also able to learn fuzzy sets and fuzzy rules [Nauck and Kruse, 1993, Nauck and Kruse, 1994], but it uses a reinforcement learning algorithm. This model is described in a paper later in this book.

4 Applicational and Semantical Aspects

If the application of a neuro-fuzzy model is considered, the semantical aspects of the underlying models should be known. Some of the models described here change their parameters in a way which does not allow an interpretation in terms of a fuzzy controller once the learning process has caused adaptions.

The weighting of rules is sometimes interpreted as "importance" of a rule, but it cannot really be viewed this way. If a rule is less important, one usually means something like that it is only seldom applicable, or that it is not harmful if the rule is not applied, but not that its conclusion should be taken into account only to some extent. This aspect is already modelled by using fuzzy sets to describe its antecedent. Weighting a fuzzy rule is actually equivalent to changing its conclusion but often in a way that leaves the domain of the fuzzy model. Individual rule weights can cause identical linguistic values to be represented in different ways within the rule base, which is usually not desirable.

The same problem occurs when approaches like ARIC do not have mechanisms to make sure that all changes caused by the learning procedure are interpretable in terms of a fuzzy system. Interpretation in terms of the usual Mamdani or Sugeno controllers is also in danger if the evaluation of antecedents is not done by t–norms, or if specialized defuzzification strategies are used.

These problems cease to exist if interpretation is not important. If there is only a need to have a working adapative controller that can use prior knowledge, any neuro–fuzzy model may be useful. In this case even a pure neural controller should be considered. But usually an interpretable architecture is of interest so that manual changes may be possible, or for reasons of security, or to be able to use standard fuzzy hardware. Interpretable architectures also have the advantage to be usable as knowledge acquisition tools.

If it is possible that the controller learns online, either at a process model or at the process itself, hybrid neuro–fuzzy models can be used. This makes it possible to find a controller for a process that can not be controlled up to now, and they can use partial prior knowledge. These models are also interesting because they allow constant adaption during the lifecycle of the controller. Approaches like NEFCON that are able to learn both fuzzy rules and fuzzy sets, and also pay attention to semantical aspects should be preferred.

If a large number of examples describing process control is available, this data can be used with a cooperative neuro–fuzzy model to determine fuzzy sets if the fuzzy rules are known, or vice versa. In this case also other means like fuzzy clustering algorithms should be considered.

If there is a problem to find an aedequate fuzzy controller for a given process, neuro–fuzzy models should not be considered as the ultimate solution. There is not much experience with these models until today, and they are usually applied to toy problems under laboratory conditions. To understand these approaches better, generic models can help, and testbeds and benchmarks are needed to compare different neuro–fuzzy systems.

5 Conclusions

This paper wants to give an overview to the large variety of neural fuzzy systems that are known today. They are usually designed for fuzzy control applications. But in the near future other areas of application like data analysis will certainly emerge. Also the use of fuzzy methods to enhance the learning algorithms of neural networks will be of further interest. One of the other articles in this chapter, the paper by Halgamuge, Mari, and Glesner, describes how fuzzy techniques can be applied to tune the parameters of the backpropagation learning procedure for multilayer perceptrons.

The paper by Felix, Kretzberg and Wehner addresses a special part of data analysis, the analysis of images. The authors describe the application of fuzzy methods and compare them to neural networks.

The two remaining papers in this chapter describe special hybrid neuro–fuzzy models. The article of Brahim and Zell present an extension to the well known Stuttgart Neural Network Simulator (SNNS) to implement the ANFIS model. The last paper in this chapter is about the NEFCON model, and presents NEFCON–I, a simulation environment to develop neural fuzzy controllers.

References

I. Aleksander and H. Morton (1990). *An Introduction to Neural Computing*. Chapman and Hall, London.

C. von Altrock, B. Krause and H. Zimmermann (1992). *Advanced Fuzzy Logic Control Technologies in Automotive Applications*. In *Proc. IEEE Int. Conf. on Fuzzy Systems 1992*, pages 835–842, San Diego.

K. Asakawa and H. Takagi (1994). *Neural Networks in Japan*. Communications of the ACM, 37(3):106–112.

A. G. Barto, R. S. Sutton and C. W. Anderson (1983). *Neuronlike Adaptive Elements that Can Solve Difficult Learning Control Problems*. IEEE Trans. Systems, Man & Cybernetics, 13:834–846.

H. R. Berenji (1992). *A Reinforcement Learning-Based Architecture for Fuzzy Logic Control*. Int. J. Approximate Reasoning, 6:267 –292.

H. R. Berenji and P. Khedkar (1992). *Learning and Tuning Fuzzy Logic Controllers Through Reinforcements*. IEEE Trans. Neural Networks, 3:724–740.

H. Bersini, J.-P. Nordvik and A. Bonarini (1993). *A Simple Direct Adaptive Fuzzy Controller Derived from its Neural Equivalent*. In *Proc. IEEE Int. Conf. on Fuzzy Systems 1993*, pages 345–350, San Francisco.

J. C. Bezdek and S. K. Pal, eds. (1992). *Fuzzy Models for Pattern Recognition*. IEEE Press, New York.

D. Driankov, H. Hellendoorn and M. Reinfrank (1993). *An Introduction to Fuzzy Control*. Springer–Verlag, Berlin.

J.-S. R. Jang (1991). *Fuzzy Modeling Using Generalized Neural Networks and Kalman Filter Algorithm*. In *Proc. of the Ninth National Conf. on Artificial Intelligence (AAAI-91)*, pages 762–767.

J.-S. R. Jang (1992). *ANFIS: Adaptive-Network-Based Fuzzy Inference Systems*. IEEE Trans. Systems, Man & Cybernetics.

B. Kosko (1992). *Neural Networks and Fuzzy Systems. A Dynamical Systems Approach to Machine Intelligence*. Prentice-Hall, Englewood Cliffs.

H. Narazaki and A. L. Ralescu (1991). *A Synthesis Method for Multi-Layered Neural Network using Fuzzy Sets*. In *IJCAI-91: Workshop on Fuzzy Logic in Artificial Intelligence*, pages 54–66, Sydney.

D. Nauck, F. Klawonn and R. Kruse (1993). *Combining Neural Networks and Fuzzy Controllers*. In E. P. Klement and W. Slany, eds.: *Fuzzy Logic in Artificial Intelligence (FLAI93)*, pages 35–46, Berlin. Springer–Verlag.

D. Nauck, F. Klawonn and K. Rudolf (1994). *Neuronale Netze und Fuzzy–Systeme (Neural Networks and Fuzzy Systems, in German)*. Vieweg–Verlag, Braunschweig.

D. Nauck and R. Kruse (1992). *Interpreting Changes in the Fuzzy Sets of a Self-Adaptive Neural Fuzzy Controller*. In *Proc. Second Int. Workshop on Industrial Applications of Fuzzy Control and Intelligent Systems (IFIS'92)*, pages 146–152, College Station, Texas.

D. Nauck and R. Kruse (1993). *A Fuzzy Neural Network Learning Fuzzy Control Rules and Membership Functions by Fuzzy Error Backpropagation*. In *Proc. IEEE Int. Conf. on Neural Networks 1993*, pages 1022–1027, San Francisco.

D. Nauck and R. Kruse (1994). *NEFCON–I: An X–Window based Simulator for Neural Fuzzy Controllers*. In *Proc. IEEE Int. Conf. Neural Networks 1994 at IEEE WCCI'94*, Orlando. To be published.

H. Nomura, I. Hayashi and N. Wakami (1992). *A Learning Method of Fuzzy Inference Rules by Descent Method*. In *Proc. IEEE Int. Conf. on Fuzzy Systems 1992*, pages 203–210, San Diego.

A. Nowé and R. Vepa (1993). *A Reinforcement Learning Algorithm based on 'Safety'*. In E. P. Klement and W. Slany, eds.: *Fuzzy Logic in Artificial Intelligence (FLAI93)*, pages 47–58, Berlin. Springer-Verlag.

W. Pedrycz and W. C. Card (1992). *Linguistic Interpretation of Self-Organizing Maps*. In *Proc. IEEE Int. Conf. on Fuzzy Systems 1992*, pages 371–378, San Diego.

T. J. Procyk and E. H. Mamdani (1979). *A Linguistic Self-Organizing Process Controller*. Automatica, 15:15–30.

W. Z. Qiao, W. P. Zhuang, T. H. Heng and S. S. Shan (1992). *A Rule Self-Regulating Fuzzy Controller*. Fuzzy Sets and Systems, 47:13–21.

R. Rojas (1993). *Theorie der Neuronalen Netze: Eine systematische Einführung*. Springer–Verlag, Berlin.

S. Shao (1988). *Fuzzy Self-Organizing Controller and its Application for Dynamic Processes*. Fuzzy Sets and Systems, 26:151–164.

P. Simpson (1992a). *Fuzzy Min-Max Neural Networks – Part 1: Classification*. IEEE Trans. Neural Networks, 3:776–786.

P. Simpson (1992b). *Fuzzy Min-Max Neural Networks – Part 2: Clustering*. IEEE Trans. Fuzzy Systems, 1:32–45.

S. M. Sulzberger, N. N. Tschichold-Gürman and S. J. Vestli (1993). *FUN: Optimization of Fuzzy Rule Based Systems using Neural Networks*. In *Proc. IEEE Int. Conf. on Neural Networks 1993*, pages 312–316, San Francisco.

H. Takagi and I. Hayashi (1991). *NN-Driven Fuzzy Reasoning*. Int. J. Approximate Reasoning, 5:191–212.

D. A. White and D. A. Sofge, eds. (1992). *Handbook of Intelligent Control. Neural, Fuzzy, and Adaptive Approaches*. Van Nostrand Reinhold, New York.

T. Yamaguchi, K. Goto, T. Takagi, K. Doya and T. Mita (1992). *Intelligent Control of a Flying Vehicle using Fuzzy Associative Memory System*. In *Proc. IEEE Int. Conf. on Fuzzy Systems 1992*, pages 1139–1149.

3.2. Image Analysis based on Fuzzy Similarities

Rudolf Felix, Thomas Kretzberg, Martin Wehner

Abstract

The so-called qualitative and structural fuzzy image analysis are desribed. For image data represented as fuzzy sets a special notion of similarity - the analogy - is used in order to compare the images. The notion of analogy has been previously applied in fuzzy decision making. Subsequently the application examples are given. Finally a comparison is made with neural networks in the field of character recognition.

1 Introduction

In the field of pattern recognition the syntactic and the decision-making-oriented approach are distinguished [16]. The syntactic approach considers the pattern recognition as solving of the word problem for formal languages. The decision-making-oriented approach uses weighted sums of image characteristics, the so-called utility functions.

Especially the industrial application of the approaches is almost impossible because of the high computational complexity of the analysis algorithms. For instance, the computational complexity of the word problems for context-free languages is $0\ (n^3)$. Since many pattern recognition problems can only be described in a context-sensitive way and the complexity of the analysis is $0\ (2^n)$, the approach is not acceptable for real world problems.

Using utility functions, the problem is to determine adequately the weights. In the case that there are interdependences between the different characteristics aggregated in the sum, the determination of the weights is of $O\,(2^n)$ similarly to the context-sensitive syntax analysis.

In [2] and [3] an approach is presented which analyzes the interdependences between characteristics of decision situations expressed as decision goals. The approach is based on different fuzzy relations describing relationships between decision goals. Based on the relationships the decision making problems are solved in polynomial computational time.

Since the decision making approach mentioned above is an altermative to decision making methods using utility functions [4], the idea to compare images using the same fuzzy relations, now applied to image characteristics instead of goals, seems to be promissing.

2 Basic Definitions

Let C be a non-empty, finite set of image characteristics $C := \{c_1, \ldots, c_n\}$, $i \in \{1, \ldots, n\}$. For every image O two fuzzy sets are defined as follows:

$$\forall_i : \mu_S{}^O(c_i) := \begin{cases} \delta, & \text{if the presence of characteristic } c_i \text{ is observed with intensity } \delta \\ 0 & \text{else} \end{cases}$$

(1)

$$\forall_i : \mu_S{}^O(c_i) := \begin{cases} \delta, & \text{if the non-presence of characteristic } c_i \text{ is observed with intensity } \delta \\ 0 & \text{else} \end{cases}$$

(2)

Every image O can be represented as two fuzzy sets S and D. The fuzzy set S describes the degree of presence of the characteristics. The fuzzy set D describes the degree of their non-presence.

Starting with such a representation of an image, two images can be compared based on fuzzy inclusions and non-inclusions between the sets S and D of the corresponding images.

Let $\mu_{\subset} : \mu^{O_1} \times \mu^{O_2} \to [0,1]$ (3)

be an inclusion relation for $O_1 \subset O_2$.

The non-inclusion $\mu_{\not\subset}$ for $O_1 \not\subset O_2$ is defined as $\mu_{\not\subset} := 1 - \mu_{\subset}$. (4)

Based on the introduced definitions the notion of similarity called "analogy" is defined as follows:

$$\mu_{analogy}(O_1,O_2) := \min\left(\mu_S^{O_1} \subset \mu_S^{O_2}, \mu_S^{O_1} \not\subset \mu_D^{O_2}, \mu_S^{O_2} \not\subset \mu_D^{O_1}, \mu_D^{O_1} \subset \mu_D^{O_2}\right)$$

where O_1 and O_2 are two images to be compared.

Note that the analogy is not a similarity relation in the sense of [5], [6], [7], [8], [9], since it is not symmetric.

3 Determination of S and D

In order to describe in which way the fuzzy sets S and D are determined let us consider grey level images as an example.

In order to perform the comparision, the image is partitioned into a grid of sectors. For each sector an average of grey levels is used as characteristic in the sense of the definition of μ_S and μ_D. A characteristic c_i is completely absent if the average grey level is white. Using a normalization of values between black and white the value of μ_S^O and μ_D^O for every sector can be calculated. The image O is then represented by the two fuzzy sets S and D.

In a similar way additional characteristics, for example the number of grey level changes, can be considered.

Note that the computational complexity of the comparison of images based on the notion of analogy is $O(n)$ where n is the number of characteristics (sectors) under consideration.

4 Qualitative and Structural Image Analysis

One of the aspects when determining the sets S and D is the granularity of the grid structure which of course implies the number of sectors and therefore the number of characteristics to be considered when representing the image.

The higher the number of sectors, the more details of the image are represented explicitely by S and D. Since the location of the sectors within the image can be expressed by indexing, structural information of the images is represented explicitely, too. Therefore, with an increasing number of sectors the analysis of the image becomes increasingly structural

The lower the number of sectors the less structural information is represented explicitely and the more qualitative becomes the analysis of the image. The term "qualitative" refers to image analysis which does not concern information about the location of the characteristics of the image. Qualitative analysis refers rather to the presence or absence of image characteristics, to their distribution inside of the image and to their intensity.

In the extreme case, each sector corresponds to one pixel in case of structural analysis. The extreme case of qualitative analysis is reached when the whole image is contained in only one sector.

5 Application Fields

Both the structural and the qualitative image analysis have already been applied in numerous analyzing tasks. The possibility to analyze contours of any kind opens a variety of applications whenever information refering to contours is relevant. For example analysis of the quality of punched work pieces, quality analysis of casings, quality of robot controlled placement and positioning actions in the process automation [11] are examples of successful applications (see Fig. 1, 2).

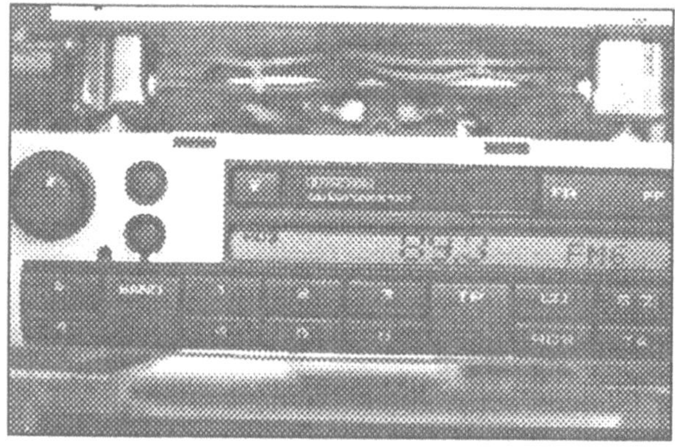

Fig.1

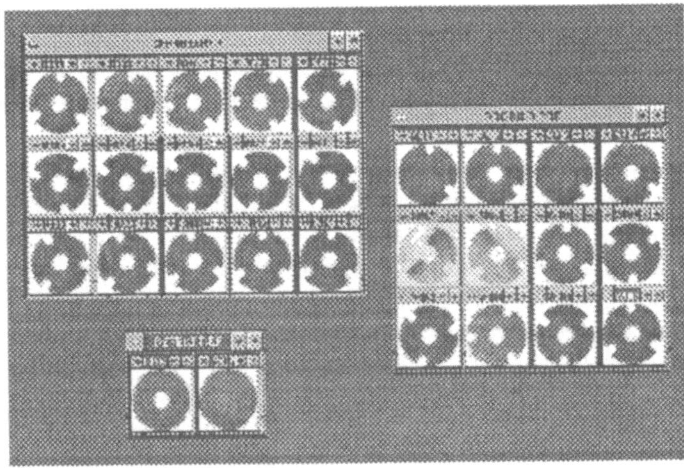

Fig.2

Qualitative image analysis has been applied in the field of surface classification [1], for example in the field of soot dispersion tests used in order to classify the quality of gum workpieces used in the automotive industry (see Fig. 3).

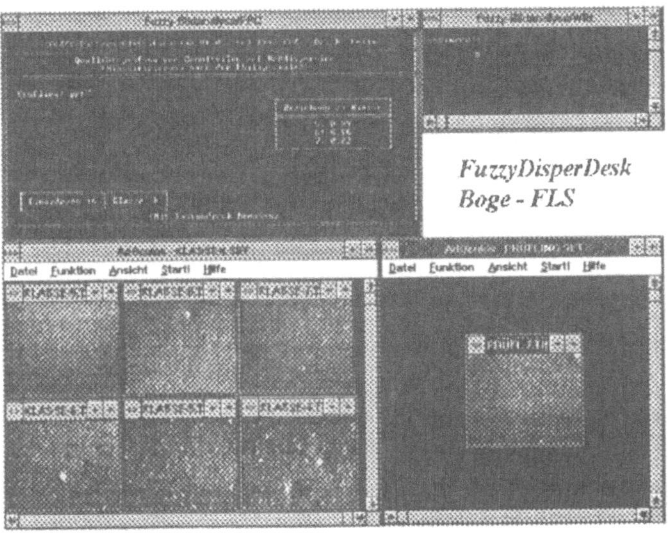

Fig. 3

6 Comparison with Neural Networks

The image analysis method presented in previous sections can be used in application fields similar to neural networks. In order to compare the performance of the method presented here with solutions based on neural networks an example application has been implemented using both methods, and the results have been compared.

The application refers to recognition of handwritten capital characters. Fig. 4 shows examples of such characters. The comparision of the results [12] shows that the performance of both methods is similar.

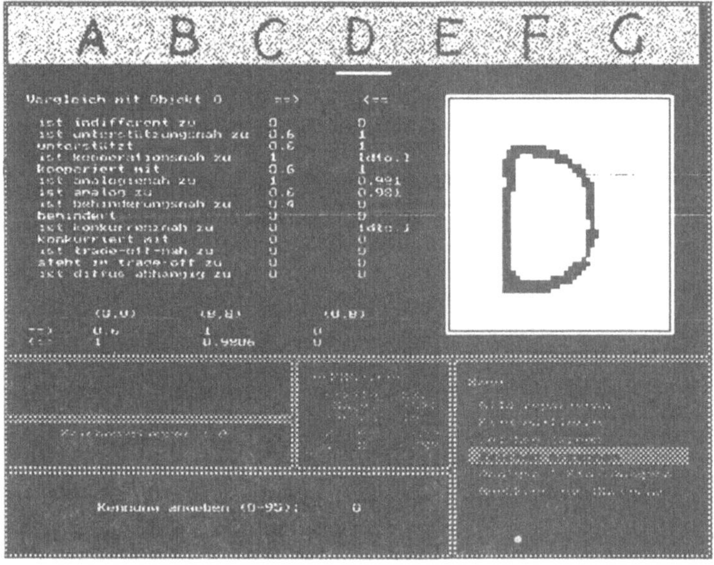

Fig.4

The advantage of the method based on the notion of analogy is the fact that the set of characters to be required can be extended incrementally in $O(n)$ computational time, where n is the number of sectors considered. For this, the only operation to be performed is to compute the sets S and D for the new character . Once the sets S and D for the new character are determined, they can be attached to the set of the old characters and the recognition can immediately be performed on the extended set of characters. This holds independent of the number of characters under consideration.

7 Summary

A method for image analysis based on a notion of fuzzy similarity has been presented. Analogy is a fuzzy relation which is used to compare images represented as fuzzy sets. In contrast to other approaches, both the presence of image characteristics (for example grey level information) and their absence are used to represent an image. Based on the notion of analogy it is shown how structural and qualitative image analysis can be performed in linear computational time. Some application examples and a comparision with neural networks based approaches are indicated. The comparision refers to the field of recognition of hand-written capital characters. It shows that the method based on the notion of analogy performs similarly to neural networks, but is able to increase incrementally the number of characters under consideration.

8 References

[1] Felix, R.; Reddig, S.: "Qualitative Pattern Analysis for Industrial Quality Assurance". *Proceedings of the 2nd IEEE Conference on Fuzzy Systems Vol I.* San Francisco, USA (March 28–April 1, 1993): 204–206.

[2] Felix, R.: "Entscheidungen bei qualitativen Zielen", Universität Dortmund, Fachbereich Informatik, Forschungsbericht Nr. 412, 1992

[3] Felix, R.: "Multiple Attribute Decision Making based on Fuzzy Relationships between Objectives", Proceedings of the 2nd International Conference on Fuzzy Logic and Neural Networks, Iizuka, Japan, 1992

[4] Felix, R.: "Goal-oriented Selection of Alternatives Based on Fuzzy Sets", IPMU Information Processing and Management of Uncertainty in Knowledge-based Systems, Paris, France, 1990

[5] Zadeh, L.A.: "Similarity Relations and Fuzzy Orderings", Yager, Ovchinnikov, Tong, Nguyen, FUZZY SETS AND APPLICATIONS, Selected Papers by L.A. Zadeh, Wiley-Interscience, 1987

[6] Ovchinnikov, S.: "Similarity relations, fuzzy partitions, and fuzzy orderings", Fuzzy Sets and Systems vol. 40, 1991, pp. 107-126, North Holland

[7] Ruspini, E.H.: "On the Semantics of Fuzzy Logic", International Journal of Approximate Reasoning, vol. 5, Numberl, 1991, pp. 45 - 88

[8] Ruspini, E.H.: "ON TRUTH, UTILITY, and SIMILARITY", Fuzzy Engineering toward Human Friendly Systems, vol. 1, Nov. 1991, pp. 42 - 50

[9] Calvo, T.: "On fuzzy similarity relations", Fuzzy Sets and Systems vol. 47, 1992, pp. 121-123, North Holland

[10] Gonzales, R. C.;Thomason, M. G.: "Syntactic Pattern Recognition - An Introduction". *Addison-Wesley*. 1978.

[11] Felix, R.; Kretzberg, T.: "Qualitative und strukturelle Fuzzy Bildanalyse". In: Reusch, B (Editor): *Fuzzy Logic - Theorie und Praxis*. 3. Dortmunder Fuzzy Tage, Dortmund, 7.-9. Juni 1993. *Springer*. 1993.

[12] Felix, R., Kretzberg, T.; Wehner, M.: "Fuzzy Entscheidungsunterstützung und Bildanalyse", Tagungsband des Workshops *Fuzzy-Systeme '93 - Management unsicherer Informationen*, Braunschweig, 1993.

3.3. ANFIS-SNNS: Adaptive Network Fuzzy Inference System in the Stuttgart Neural Network Simulator

Kais Brahim, Andreas Zell

Abstract

In this paper the Neuro-Fuzzy system ANFIS (*Adaptive Network Fuzzy Inference System*) and its integration in the Stuttgart Neural Network Simulator (SNNS) is described. The rule-based knowledge base of a fuzzy system is directly mapped to the network structure of a neural network. With a hybrid learning algorithm the system adapts itself to the environment by using examples to optimize the rules. The structured network architecture also gives the possibility to extract the optimized fuzzy rules from the network after training.

1 Introduction

ANFIS (*Adaptive Network Fuzzy Inference System*) [Jan92a] is a Neuro-Fuzzy system which realizes an interesting combination of fuzzy logic with neural networks. An existing knowledge base of fuzzy IF-THEN-rules, which may be generated, for example, by a fuzzy modeling environment (FME) [Bra93], is directly mapped to the structure of a neural network. The neural network may then be trained with conventional learning algorithms like backpropagation [RM86] or with a new hybrid learning algorithm. In both methods additional training examples are used to optimize the existing fuzzy rules. The special network architecture permits to extract the optimized fuzzy rules from the trained network. Coarsely specified rules are sufficient to initialize the network, since they are optimized with the training data during learning.

This methodology simplifies modeling of systems with fuzzy logic because it allows an automatic adaptation to changing characteristics of the modeled process. In this way a rapid incremental system design is possible in which first the existing knowledge is coded in fuzzy rules and later these rules are refined and adapted to the changing process.

Different fuzzy inference types (Tsukamoto, Max-Min, and Takagi-Sugeno) are mapped to three different network architectures (ANFIS-1, ANFIS-2 and ANFIS-3). The main difference lies in the specification of the conclusions of the IF-THEN-rules.

2 Description of the ANFIS Architectures

The ANFIS architectures are special multi-layer feedforward networks consisting of adaptive and non-adaptive units (cells, artificial neurons), which communicate by directed links (see Fig. 1). Adaptive elements possess parameters which are optimized during learning.

Viewing Fig. 1 we note that layer two consists of adaptive units, where each unit possesses a number of premise parameters ($\in \mathcal{S}_1$) to define its membership function. Units in the layer preceding the last layer are also adaptive. Their parameters define the set of conclusion parameters $\mathcal{S}_2$.

2.1 ANFIS-1

These networks (see Fig. 1) realize a fuzzy inference after Tsukamoto. For simplicity, in the conclusion part (layer 5) a linear approximation of the monotonously increasing membership function is performed.

The example network in Fig. 1 contains the following rule base:

Rule 1: IF x IS A_1 AND y IS B_1 THEN $o_1^5 = f_1(o_1^3, o_1^4)$
Rule 2: IF x IS A_2 AND y IS B_2 THEN $o_2^5 = f_2(o_2^3, o_2^4)$

2.2 ANFIS-2

These networks realize a fuzzy inference after the Max-Min method. One network layer is omitted in comparison to ANFIS-1.

Rule 1: IF x IS A_1 AND y IS B_1 THEN z_1 IS C_1
Rule 2: IF x IS A_2 AND y IS B_2 THEN z_2 IS C_2

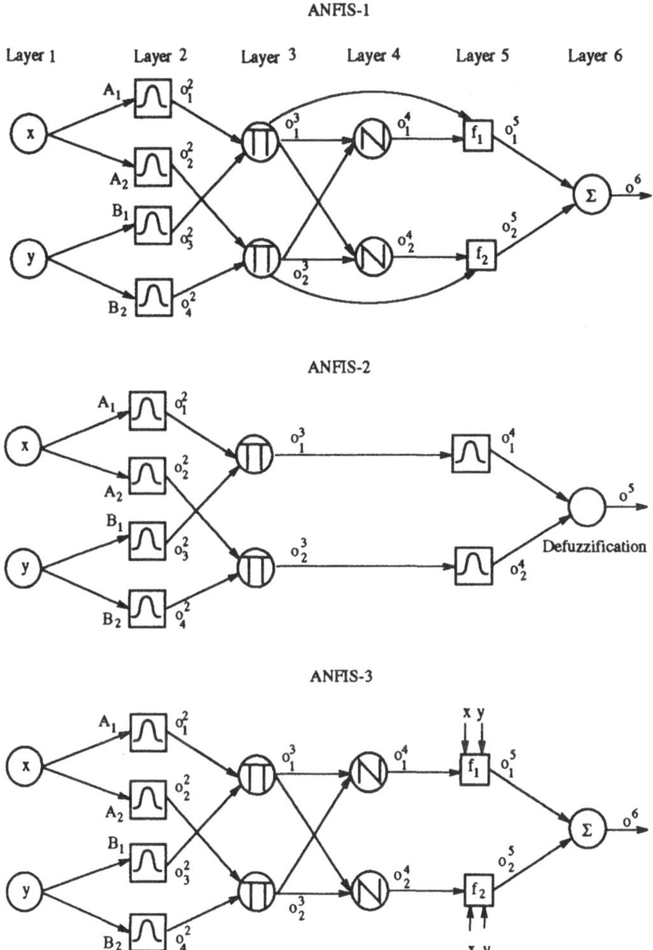

Figure 1: Example networks for the three ANFIS network types

The adaptive cells in layer 4 perform their inference after the Max-Min inference method. The output unit computes a defuzzification. This network type realizes an inference which is more complicated than types 1 or 3. This considerably slows down learning because of the more difficult parameter adjustment. In this method parameters may only be adjusted by a gradient descent learning. For this reason in our neural network fuzzy logic simulation environment no neural network optimization method has been implemented for this type of inference yet.

2.3 ANFIS-3

This network realizes a Takagi-Sugeno inference which is more powerful than the inference of type 1.

Rule 1: IF x IS A_1 AND y IS B_1 THEN $o_1^5 = f_1(x,y) = p_1x + q_1y + r_1$
Rule 2: IF x IS A_2 AND y IS B_2 THEN $o_2^5 = f_2(x,y) = p_2x + q_2y + r_2$

where unit i in layer 5 contains the conclusion-parameters $\{p_i, q_i, r_i\}$.

3 ANFIS Learning Algorithms

Learning algorithms for ANFIS networks may be based on gradient descent algorithms like backpropagation [RM86]. Since this class of learning algorithms is rather slow and its convergence is restricted by the existence of local minima, a hybrid learning algorithm is presented which may be applied to ANFIS networks and other feedforward-networks of similar structure. Two learning algorithms (*hybrid offline* and *hybrid online*) adjust the parameter set $\mathcal{S} = \mathcal{S}_1 \uplus \mathcal{S}_2$. A learning cycle consists of a *forward-* and a *backward pass* (see table 1).

	Forward Pass	Backward Pass
Premiss parameter $\mathcal{S}_1$	unmodified	gradient descent
Conclusions parameter $\mathcal{S}_2$	LSE	unmodified

Table 1: The ANFIS hybrid learning rule.

3.1 The Forward Pass

To simplify the description it is assumed that the ANFIS network has only one output unit. The algorithm may easily be generalized to several output units. The output *out* is generated as follows:

$$
\begin{aligned}
out &= F(\vec{I}, \mathcal{S}) && (1)\\
H(out) &= H \circ F(\vec{I}, \mathcal{S}) && (2)\\
B &= AX && (3)\\
(A^T A)^{-1} A^T B &= X^* && (4)
\end{aligned}
$$

Here $\vec{I}$ is the vector of all input variables, $\mathcal{S}$ the set of all parameters of all cells. There exists a function H such that $H \circ F$ is linear in the premise parameters $\mathcal{S}_2$. P examples change equation (2) into the matrix equation (3), where the unknown vector X contains the elements of $\mathcal{S}_2$.

The number of elements in $\mathcal{S}_2$ (=M) is usually much smaller than the number of training data (P), thus equation (3) has no exact solution. The LSE algorithm minimizes the error $||AX - B||^2$ by approximating X with X^* (4). The direct computation of X^* involves a time consuming computation of the matrix inverse $(A^T A)^{-1}$ and requires that the matrix $(A^T A)$ is non-singular and well-conditioned. Therefore an iterative method to compute X^* is used [Str90]:

$$
\begin{aligned}
S_{i+1} &= S_i - \frac{S_i a_{i+1} a_{i+1}^T S_i}{1 + a_{i+1} S_i a_{i+1}} && (5)\\
X_{i+1} &= X_i + S_{i+1} a_{i+1} (b_{i+1}^T - a_{i+1}^T X_i) && (6)
\end{aligned}
$$

Here $i = 0, 1, \ldots, P-1$, P is the number of training patterns, $X_0 = \vec{0}$, $S_0 = \gamma I$, where γ is a large number, a_i^T is the i-th line of the matrix A, b_i^T the i-th element of the vector B, $X^* = X_P$. The above equations may be generalized to networks with multiple output units by making b_i^T the i-th row of the matrix B.

3.2 The Backward Pass

The error E_p is propagated back to modify the premise parameters $\mathcal{S}_1$. For each pattern pair p the error derivatives $\frac{\delta E_p}{\delta o}$ are calculated for every cell o. The error derivative $\frac{\delta E_p}{\delta o}$ at the output layer L is equal to $-2(t_p - o)$.

For an internal cell i in layer k the error derivative may be computed with the following chain rule: $\frac{\delta E_p}{\delta o_{ip}^k} = \sum_{m=1}^{\#(k+1)} \frac{\delta E_p}{\delta o_{mp}^{k+1}} \frac{\delta o_{mp}^{k+1}}{\delta o_{ip}^k}$, where $1 \leq k \leq L-1$ (L = number of layers).

With this chain rule the error derivatives of the output cells as well as internal (hidden) cells may be computed, the latter based on the error derivatives of succeeding layers.

3.3 Simplified Fuzzy IF-THEN-Rules

In ANFIS-1 the activation (membership) functions of layer 5 cells are restricted to monotonously increasing functions, which may not model linguistic variables with convex membership functions. In ANFIS-2 a defuzzification is necessary, and the systematic adjustment of parameters is time-consuming. In ANFIS-3 cells of layer 5 may not be connected with linguistic variables. To alleviate these shortcomings, simplified IF-THEN-rules of the following form are introduced:

$$\text{IF } x \text{ IS } A \text{ AND } y \text{ IS } B \text{ THEN } z \text{ IS } d$$

where d is a constant. This class of rules may be realized with all three ANFIS types. Despite the restriction of the output to the constant d these types of ANFIS networks may still possess the capability to approximate non-linear functions from example data points. [Jan92a] (Stone-Weierstra"s Theorem).

This simplified ANFIS architecture furthermore performs a type of inference which is functionally equivalent to Radial Basis Functions (RBF) [Jan92b] [Vog92]. The equivalence only holds when gaussian membership functions are used.

The hybrid learning algorithm may also be applied directly to radial basis functions. Similarly, algorithms used in the theory of radial basis functions to find optimal RBF parameters may be used in ANFIS.

4 Integration of ANFIS in SNNS

The ANFIS networks and learning algorithms described above were integrated into a version of the Stuttgart Neural Network Simulator (SNNS). SNNS is an efficient simulator of neural networks for Unix workstations with a sophisticated graphical user interface under X-Windows [ZMH+92]. Meanwhile it comprises approximately 20 different network types and learning algorithms and is used in some 800 installations world-wide. SNNS (without ANFIS) is available for research purposes free of charge via anonymous ftp from the University of Stuttgart, Institute for Parallel and Distributed High-Performance Systems (IPVR).

The integration of ANFIS in SNNS consisted of the following items:

- Creation of ANFIS networks by a network conversion tool, which compiles the rules from a fuzzy modeling environment (FME).
- Integration of different transfer functions in SNNS.
- Development of initialization routines with LSE (*least square estimation*) to start the learning algorithms.
- Integration of the hybrid learning algorithms (online and offline). This additional training yielded resulted in improved performance for the ANFIS-1 and ANFIS-3 network types.

4.1 Extensions to the SNNS kernel

Several functions to integrate ANFIS networks in SNNS were implemented.

1. A special conversion tool converts rules from a fuzzy modeling environment FME [Bra93] to ANFIS neural networks in SNNS format. Thus tested and debugged fuzzy systems may be used as a starting point which considerably shortens network training time. Furthermore the heterogeneous network structure may be generated rapidly and without errors.
2. The SNNS unit data structures were extended to include the premise and conclusions parameters necessary in ANFIS. Each cell has the parameters a, b, c defining the shape of the membership function and a pointer to an additional array of conclusion parameters which is used by the LSE algorithm.
3. For loading and writing of ANFIS networks two new I/O functions were implemented, which additionally set premise and conclusion parameters and which write them to file together with the other network parameters. The SNNS network format was extended by a new section to define the parameters of the fuzzy logic units.
4. Different matrix operations were added which were necessary to implement the iterative LSE algorithm.

5. In ANFIS-1 and ANFIS-3 two initialization algorithms to determine conclusion parameters with the LSE algorithm were implemented.

6. A number of new transfer functions like T-Norm and inference operators for the ANFIS models were integrated in SNNS.

7. ANFIS learning algorithms for ANFIS-1 and ANFIS-3 in online and batch mode were integrated in SNNS

The tests performed with the resulting ANFIS networks have shown that training in online mode is affected by numerical instabilities. Training in batch mode, on the other hand, is more stable.

4.2 Application Example

The use of ANFIS networks is demonstrated with the following example [NP90] [Jan92a] in which the neural network used in this task (1-20-10-1) is replaced by an ANFIS-3 network. The task is the identification of a non-linear component f of a controller with ANFIS-3. The behavior of the system is described mathematically by the following differential equation:

$$y(t+1) = 0.3y(t) + 0.6y(t-1) + f(u(t)) \qquad (7)$$

where t is the time index, $u(t)$ the input, $y(t)$ the output and $f()$ the function which is to be approximated. The Training in ANFIS used 250 training pattern pairs (u, f), which were created by the following function:

$$u(t) = \sin(2\pi t/250), 1 \leq t \leq 250 \qquad (8)$$
$$f(u) = 0.3\sin(\pi u) + 0.6\sin(3\pi u) + 0.5\sin(5\pi u) \qquad (9)$$

The ANFIS-3 network used in online learning consists of 7 units in layer 2, i.e. 7 membership functions for the input x which generate 7 rules. In total the hybrid learning algorithm fits 21 premise parameters and 14 conclusion parameters to the training data. The network is described in Fig. 2. With seven membership functions hardly any change of the premise parameters takes place. Most changes are performed at the conclusion parameters. With as few as three rules the system shows the first small differences to the behavior or the physical system. Table

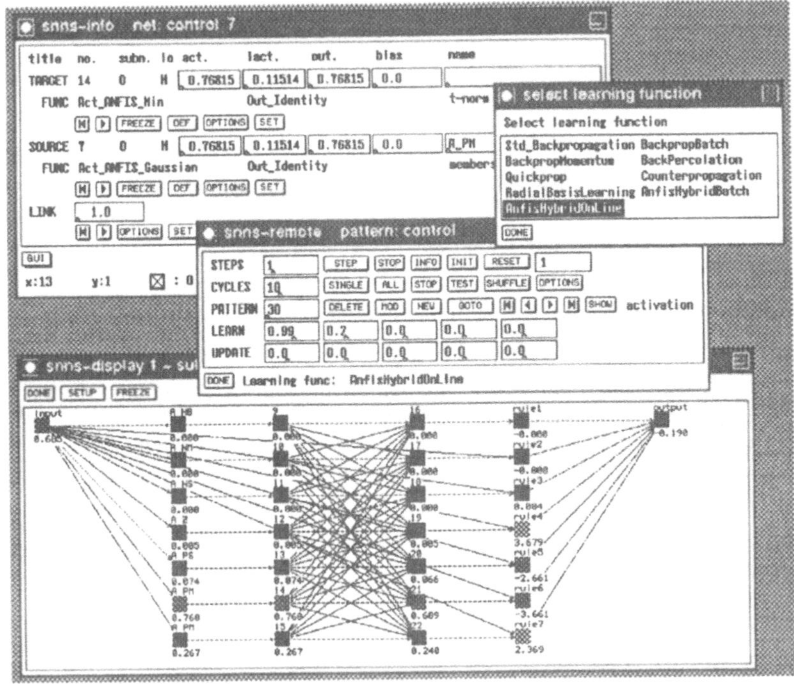

Figure 2: ANFIS network in SNNS with 7 membership functions

Method	Parameters	Learning cycles
Neural Network (1-20-10-1)	261	50000
ANFIS-3 (online)	35	250

Table 2: Performance comparison of ANFIS with neural network given in [NP90]

2 summarizes the performance comparison of the ANFIS fuzzy rules or ANFIS network with the standard neural network given in [NP90].

The number of network parameters may be reduced further by using fewer membership functions and training the network in offline (batch)-mode. Networks with 5, 4 and even 3 membership functions (rules) have shown good results with as few as 50 training cycles.

5 Summary

The implementation of *Adaptive Network Fuzzy Inference Systems* (ANFIS) in SNNS permits an optimization of a fuzzy system based on training data extracted from the actual working environment. Fuzzy systems with were created, tested and saved as rules in a fuzzy modeling environment (FME) are automatically mapped to three different types of ANFIS networks, which differ by their network structure and their inference method. With hybrid learning algorithms, which consist of a gradient descent algorithm combined with the method of least square estimation (LSE), the parameters of the network are adapted to the training data of the environment. The ANFIS architecture also gives the possibility to directly optimize a knowledge base in the form of IF-THEN-rules and to extract it as rules after optimization. The capabilities of ANFIS in comparison with standard feedforward neural networks in online identification of a controller were shown with an example.

References

[Bra93] K. Brahim. Methoden zur Kombination von Fuzzy Logik und Neuronalen Netzen. Diplomarbeit 978, IPVR, University of Stuttgart, 1993.

[Jan92a] J.-S. Roger Jang. ANFIS: Adaptive-Network-Based Fuzzy Inference Systems. *IEEE Transactions on Systems, Man & Cybernetics*, 1992.

[Jan92b] J.-S. Roger Jang. Functional Equivalence between Radial Basis Function Networks and Fuzzy Inference Systems. *IEEE Transactions on Neural Networks*, 1992.

[NP90] K.S. Narendra, K. Parthsarathy. Identification and Control of Dynamical Systems Using Neural Networks. *IEEE Transactions on Neural Networks*, 1(1):4–27, 1990.

[RM86] E. Rumelhart, J.L. McClelland. *Parallel Distributed Processing: Explorations in the Microstructure of Cognition.*, Band I, II. MIT Press,Cambridge, Massachusetts, London, England, 1986.

[Str90] P. Strobach. *Linear Prediction Theory : A Mathematical Basis for Adaptive Systems.* Springer-Verlag, 1990.

[Vog92] M. Vogt. Implementierung und Anwendung von "'Generalized Radial Basis Functions"' in einem Simulator neuronaler Netze. Diplomarbeit 875, IPVR, Universität Stuttgart, 1992.

[ZMH+92] A. Zell, N. Mache, R. H"ubner, M. Schmalzl, T. Sommer, G. Mamier, M. Vogt, T. Korb. *SNNS User Manual, Version 2.1.* IPVR, Universit"at Stuttgart, 1992.

3.4. Fast Perceptron Learning by Fuzzy Controlled Dynamic Adaptation of Network Parameters

Saman K. Halgamuge, Andreas Mari and Manfred Glesner

Abstract

Application of fuzzy control for obtaining better performance from conventional neural networks is a new area in the field of fuzzy-neural combined systems. Conventional backpropagation algorithm for example can be improved by changing network parameters, based on the empirical knowledge gained by the user. This manual adaptation is effectively replaced by a fuzzy controller that contains the a priori knowledge in form from membership functions and rules. The implemented modified backpropagation algorithm with a fuzzy controller for dynamic adaptation of network parameters is tested with a benchmark data set and two real world problems.

1 Introduction

After several applications in Japan based on theoretical foundations and fundamental techniques developed in the last few decades, a growing up of huge interest in fuzzy systems can be seen in Europe.

Since neural networks (NN) belong to the same category of model-free estimators as fuzzy systems (FS), many researchers are successful in combining them. Fusion techniques for neural networks and fuzzy systems can be divided into three categories. Neural networks can be applied to generate and tune FS (e.g. [Kosko, 1992], [Halgamuge and Glesner, 1994b]), efficiency of neural networks can be improved by using FS (e.g. [Arabshahi *et al.*, 1992], [Xu *et al.*, 1992]), and neural and fuzzy systems can be cascaded.

This paper belongs to the second category, and reports a successful attempt of using fuzzy control in improving perceptron learning. After a brief explanation on theoretical background of the backpropagation algorithm [Rumelhart and McClelland, 1986] giving more emphasis on speed influencing parameters, the authors present the fuzzy controlled dynamic adaptation method. Finally the obtained results are given, and future work is discussed.

2 Speed Influencing Parameters in the Backpropagation Algorithm

The originally proposed formula ([Hertz *et al.*, 1991]) for computation of the weight updates in a multilayer perceptron is

$$w_{ij}(t+1) = -\eta \frac{\partial E}{\partial w_{ij}} + w_{ij}(t) \tag{1}$$

where w_{ij} stands for the weight value from the jth neuron to the ith neuron, η for the *learning rate* and t denotes the iteration count. According to the above equation the computation of the partial derivatives of an error function E is used to measure how well the network is able to classify presented patterns. Several such functions have been proposed. The most commonly used (and therefore also here implemented) is the *sum of squares* function

$$E[\mathbf{w}] = \frac{1}{2}(\sum_{\mu}\sum_{i}[\zeta_i^{\mu} - O_i^{\mu}]^2) \tag{2}$$

where ζ is used for the desired and O for the actual output. The double sum is necessary since the perceptron used in the described research work was trained in batch mode and μ denotes the presented patterns.

The output of each neuron has to be defined by another mathematical function, the so called *activation function*. Again several functions have been employed and one of the most important is the *sigmoid function*

$$g(h) = \frac{1}{1 + \exp[-\beta h]} \tag{3}$$

where h is the sum of the weighted inputs to the neuron and β the *steepness parameter*.

The computation method mainly defined by (1) is known as *gradient descent* and its properties have been documented by many authors ([Hertz *et al.*, 1991], [Rumelhart and McClelland, 1986]). Since the method is inherently slow, great efforts have been taken to improve the speed by introducing new parameters (*momentum* [Plaut *et al.*, 1986], *adaptive parameters* [Jacobs, 1988], *individual learning rate* [Silva and Almeida, 1990] etc.).

However, there are only a few parameters allowing for improvement and their interaction is rather complex. This is the reason why most methods, though based on the mathematical properties of the formulas, are empirically derived.

The learning rate η] scales the computed weight change and has therefore the biggest influence on learning speed. Ideally it must be large enough to ensure fast convergence towards the error minimum without overshooting it. The steepness β] determines the steepness of the activation function in the sensitive area (input between ± 6 approximately). The larger this value is, the closer the activation function gets to the step function, which means increasing insensitivity to different input values with the output becoming more polarized towards the upper and lower bound (0 and 1 respectively). The momentum parameter influences the learning rate according to the learning history and is therefore an adaptive method already.

In conventional backpropagation these parameters are constants chosen before computation starts. Adaptive methods follow the idea to change parameters from iteration to iteration by observation of one or more criteria. Commonly employed in non-fuzzy methods is the derivative of the error function with respect to the single weights. The idea is to use the change in the sign of the derivative as an indicator that the error function minimum for the weight was overshot. A previous paper [Jacobs, 1988] collected a number of heuristic rules for the adaptation of single learning rates (each weight is assigned its own learning rate) which form the basis of most adaptive approaches.

3 Fuzzy Parameter Adaptation Methods

It has already been mentioned that most of the suggested improvements are largely derived from empirical observations. This is an ideal field for application of fuzzy logic, the idea being to integrate the empirical knowledge in a fuzzy *rule base* that allows an automated adaptation of the speed influencing parameters. An additional benefit will be the fact that the derived perceptron is a self adaptive system where no particular values must be chosen prior to the start of computation. Of course there is the necessity to initialize the parameters but the choice does not effect the efficiency of the learning run.

The use of fuzzy logic for parameter adaptation has been proposed by several authors, e.g. [Arabshahi *et al.*, 1992], whose algorithm will be referred to as FA1, and [Xu *et al.*, 1992]). The method introduced in this paper (called FA2) is an extension of their research work suggesting an entirely different rule base. The major departure with this rule base is the computation of a learning rate within an (theoretically) unlimited range.

Compared to the non-fuzzy methods mentioned, the fuzzy methods proposed so far (including this paper) use only one learning rate parameter for all weights. It is therefore not useful to apply the error derivative criterion described in the previous section. Instead every approach towards learning rate adaptation must consider the change in (overall) error.

The main observations are listed and complemented by additional statements from further empirical studies made for this research to give the foundation for a rule base. It should be noted that all descriptions are qualitative at this stage. Actual quantities will be incorporated in the membership functions.

1. A high error means being far away from the minimum. Hence the learning rate should be high.

2. The change of the error (CE in short) from iteration to iteration is the most important and significant measure:

$$CE(t) = E(t) - E(t-1) \ . \tag{4}$$

   As long as it is high the learning rate can be increased quite safely. Negative CE indicates that the minimum has been passed.

3. Since the idea is to use a very large learning rate if possible it is important to know when to decrease it again as to avoid an overshot. The experiments have shown that the most reliable measure is the second change of error. Originally only the sign of this value was regarded. Positive sign means an increase in CE which in turn means it is safe to increase the learning rate. But it also proved beneficial to take the magnitude of this value into account because it contains information about the trend in CE: Even if CE remains positive for consecutive iterations, its decrease gives an early hint that the minimum is being approached. To be independent of the magnitude of CE the QCE-measure is introduced as a quotient computed as

$$QCE = CE(t)/CE(t-1) \ . \tag{5}$$

   Values of QCE smaller than 1 should lead to a decrease of the learning rate.

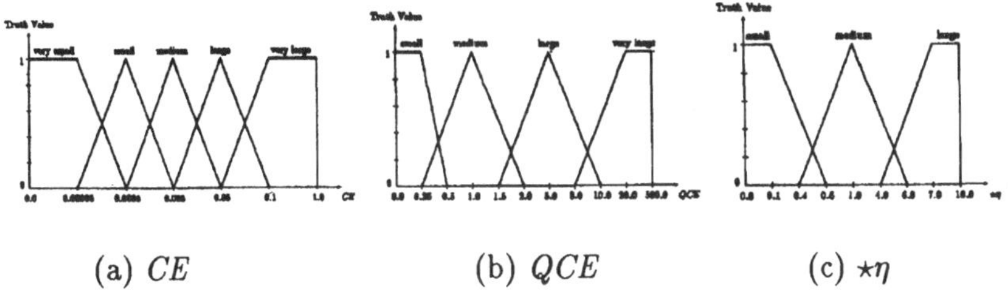

(a) CE (b) QCE (c) $\star\eta$

Figure 1: Membership Functions

CE	QCE			
	small	medium	large	very large
very small	small	small	medium	large
small	small	medium	medium	large
medium	small	medium	medium	large
large	small	medium	large	large
very large	small	medium	large	large
Output Parameter : $\star\eta$				

Table 1: Rule Base for the Learning Rate

These statements are introduced under the assumption that the error surface for each weight can be approximated by a parabola, so that an approximation of the minimum means a flattening of the curve and hence a decrease in the change of error. Now the rule base for the learning rate can be easily derived and is given in table 1.

The result of the fuzzy computation is a factor $\star\eta$ used for multiplication with the learning rate of the current step to compute the learning rate of the next step according to the formula

$$\eta(t+1) = \star\eta * \eta(t) \ . \tag{6}$$

The quantitative realization of the rule base parameters in the form of membership functions that can be implemented directly are shown in figure 1.

In [Xu *et al.*, 1992] it was shown that further speed improvement can be achieved by combining the learning rate adaptation with that of the steepness parameter. So it seemed useful to try this for the new method as well. According to the mentioned article a rule base for the steepness can be derived from the following statements:

E				
very small	small	medium	large	very large
very large	large	medium	small	very small
Output Parameter : β				

Table 2: Rule Base for the Steepness

1. If the error is large then the output of the neurons is far from the expected result. Therefore the steepness should be quite small so that the interconnections of the neurons are not driven into saturation.

2. When the error becomes smaller the steepness value should grow to adjust the weights quickly and precisely towards the correct values.

Since it was not yet possible to conduct an extensive research in how the new method of learning rate adaptation is related to changes in the steepness, the rule base and membership functions suggested in [Xu *et al.*, 1992] were changed only slightly on an intuitive basis. The modified rule base is listed in table 2.

In the actual implementation of the outlined principles it proved necessary to add some enhancements. The high learning rate will, despite the measures described, occasionally lead to an overshot. Hence only steps that decrease the error function are allowed. Otherwise the algorithm will return to the second last iteration and compute a new learning rate by simply decreasing the current one. The experiments have shown that the most reliable way to ensure a successful iteration is to use a factor that will decrease the rate by an order of magnitude.

It is also necessary to observe the learning rate to prevent it from falling below a minimum level. Otherwise the algorithm will get stuck, since no significant changes to the results can be made. So a lower bound for the learning rate is introduced and if it is reached, the learning rate is set to a fixed value instead.

Finally, when dealing with a variable steepness parameter, it is important to understand the following connection: In the update formula for the weights (1) the steepness occurs as a factor multiplied with the learning rate, in effect increasing or decreasing it depending on whether it is larger/smaller than 1 (this equation is part of the computation formula for the error derivatives; see [Hertz *et al.*, 1991]):

$$g'(h) = \beta g(1 - g) . \tag{7}$$

This is probably the main reason for the non-convergence of conventional BP if the steepness is chosen too high, since the resulting effective learning rate can easily lead to oscillations. To prevent this effect the implementation proposed in this paper uses a modified update formula:

$$w_{ij}(t+1) = -\frac{\eta}{\beta}\frac{\partial E}{\partial w_{ij}} + w_{ij}(t) \,. \tag{8}$$

4 Results

One of the main problems in comparing results is the wide variety of parameters that have to be chosen initially. This includes many that have not been considered in this research work such as *weight initialization* and *number of hidden layer neurons.*

To develop the BP-approach discussed in this paper a set of botanical data, the Iris-data of Anderson [Anderson, 1935] has been selected. The data are a set of four-dimensional vectors, each of which represents sepal length, sepal width, petal length and petal width of one of three Iris subspecies Setosa, Versicolor, and Virginica. Measurements are taken from 50 plants for each subspecies. The data set is divided into a training- and a test set, each of them consisting of 75 vectors.

The initial weight values were chosen with a randomization algorithm in the range of ± 1 but were identical for all compared methods. A two layer perceptron was implemented with 4 inputs, 3 outputs and 3 hidden layer neurons. The initial values for the crucial parameters learning rate and steepness were set to 1. Figure 2 shows the behaviour of the error over the iteration steps for three implementations to compare the two mentioned fuzzy-methods with conventional BP.

The diagram shows how the learning process benefits from the learning rate adaptation. There is a high gain of speed for FA1 already. The proposed new method FA2 performs still better, decreasing the error rapidly from the very beginning. It can be seen that the acceleration effect wears out after approximately 1000 iterations with the learning curve becoming flat. But at that point a very high classification rate is already reached. The error function is important as a means to show the overall learning speed of an algorithm on the training set. But the success of the network learning is primarily evaluated by the performance on the test set.

Two facts should be noted: A total classification success in the training set (100% corresponding to an error of 0) is not necessarily desirable

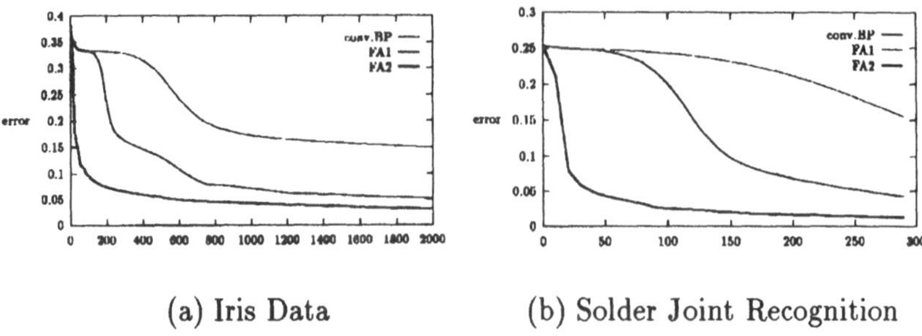

(a) Iris Data (b) Solder Joint Recognition

Figure 2: Error Function comparison for three implemented methods

since it can mean that the Network has perfectly memorized the training set but is not able to generalize. On the other hand the actual classification ratio of the test set is also dependent on the quality of the presented data patterns. For the same data set the best achieved result reported [Halgamuge *et al.*, 1993a] under another classification method is a classification ratio of 98.67% on the test data. Therefore the presented algorithm performs well. The trend that shows in the error function is confirmed by the classification ratio. The high value of 88.0% on the test set is reached by FA2 after only 400 iterations. After 2000 iterations it reaches the maximum classification rate of 90.67%. A threshold value of 0.5 is used to decide finally whether the output should be considered as 0 or 1.

There is only a very modest gain in speed when the steepness is fuzzy controlled, but it is not unlikely that further research in the connection of learning rate and steepness is able to suggest a better rule base that can improve the behaviour.

It was already mentioned that for every learning run there is a certain number of failed iterations not shown in the above statistics. The ratio of failed to successful iterations is fairly constant at around 15% which has only little effect on the performance of the algorithm.

Bearing in mind that a lower limit has been implemented for the learning rate (0.2) it can be seen that η varies over a range of five orders of magnitude. That is a considerable difference to the values that are reached with FA1 or recommended in conventional BP. They can only work because the fuzzy control is able to react immediately to the network behaviour.

Having obtained these promising results with the new method the simulations were expanded to other data sets in order to confirm the applicability to general problems. Two more data sets dealing with real world type of problems: sensor data for road surface recognition of a moving car [Halgamuge *et al.*, 1993a] and data from a laser scanner for recognition of solder joints [Halgamuge *et al.*, 1993b] are used. The results obtained under the same circumstances as for the Iris-data are analysed and validated. Figure 2 (b) shows the behaviour for solder joint recognition.

5 Discussion

The presented results show the advantage of the newly proposed method as compared to conventional BP and previously suggested methods of fuzzy adaptation of the learning rate. The results obtained with the Iris-data that were used to develop the new method are confirmed by further simulations with different data sets. So it can be safely concluded that the new method will produce good results when applied generally.

Additional benefits of the new method are the simplification in the initial choice of parameters and the fact that the computation of the adaptive parameters (namely the learning rate) is independent of the size of the network (as opposed to methods that use single learning rates for every weight). However, with ever increasing performance of parallel computing systems it would be very interesting to find out how fuzzy logic can improve on single learning rates ([Halgamuge and Glesner, 1994a]), since very good results have been reported for non-fuzzy methods applied to their adaptation ([Silva and Almeida, 1990], [Fahlman, 1988]).

Finally it must be mentioned that there is also a potential for improvement in conducting further research in the connection between learning rate and steepness that can lead to a better rule base for the steepness than the somewhat intuitive one presented here.

References

[Anderson, 1935] E. Anderson. The Irises of the Gaspe Peninsula. *Bull. Amer. Iris Soc.*, 59:2–5, 1935.

[Arabshahi *et al.*, 1992] P. Arabshahi, J. J. Choi, R. J. Marks, and T. P. Caudell. Fuzzy Control of Backpropagation. In *IEEE International Conference on Fuzzy Systems*, March 1992.

[Fahlman, 1988] S. E. Fahlman. An Empirical Study of Learning Speed in Back-Propagation Networks. In *Proceedings of the 1988 Connectionists Models Summer School*, 1988.

[Halgamuge and Glesner, 1994a] S. K. Halgamuge and M. Glesner. Fuzzy Controlled Adaptive Gradient Descent Learning Methods. In *European Congress on Fuzzy and Intelligent Technologies' 94, (accepted)*, Aachen, Germany, September 1994.

[Halgamuge and Glesner, 1994b] S. K. Halgamuge and M. Glesner. Neural Networks in Designing Fuzzy Systems for Real World Applications. *International Journal for Fuzzy Sets and Systems (in press) (Editor: H.-J. Zimmermann)*, 1994.

[Halgamuge *et al.*, 1993a] S. K. Halgamuge, H.-J. Herpel, and M. Glesner. Echtzeit Fahrbahnzustandserkennung mit Fuzzy-Neuronalen Netzen. In B. Reusch (Hrsg.), editor, *Fuzzy Logic — Theorie und Praxis*, pages 204–211. Springer, Berlin, 1993. ISBN 3-540-57524-3.

[Halgamuge *et al.*, 1993b] S. K. Halgamuge, W. Poechmueller, and M. Glesner. A Rule based Prototype System for Automatic Classification in Industrial Quality Control. In *IEEE International Conference on Neural Networks' 93*, pages 238–243, San Francisco, USA, March 1993. IEEE Service Center; Piscataway. ISBN 0-7803-0999-5.

[Hertz *et al.*, 1991] J. Hertz, A. Krogh, and R. G. Palmer. *Introduction to the Theory of Neural Computation.* Addison–Wesley, 1991.

[Jacobs, 1988] R. A. Jacobs. Increased Rates of Convergence Through Learning Rate Adaptation. *IEEE Transactions on Neural Networks*, 1988.

[Kosko, 1992] B. Kosko. *Neural Networks and Fuzzy Systems.* Prentice-Hall, USA, 1992.

[Plaut *et al.*, 1986] D. Plaut, S. Nowlan, and G. Hinton. *Report CMU-CS-86-126: Experiments on Learning in Back Propagation.* Carnegie Mellon University, 1986.

[Rumelhart and McClelland, 1986] D. E. Rumelhart and J. L. McClelland. *Parallel Distributed Processing: Explorations in the Microstructure of Cognition.* MIT Press, USA, 1986.

[Silva and Almeida, 1990] F. M. Silva and L. B. Almeida. Acceleration Techniques for the Backpropagation Algorithm. In *Lecture Notes in Computer Science*, 1990.

[Xu *et al.*, 1992] H. Y. Xu, G. Z. Wang, and C. B. Baird. A Fuzzy Neural Networks Technique with Fast Back Propagation Learning. In *International Joint Conference on Neural Networks*, June 1992.

3.5. Building Neural Fuzzy Controllers with NEFCON–I

Detlef Nauck

1 Introduction

Combinations of neural networks and fuzzy controllers that are known from recent publications are mostly *cooperative* in nature. This means a neural network is used to learn either fuzzy sets or fuzzy rules, and the results are used to build a conventional fuzzy controller. *Hybrid* approaches on the other hand try to find a new kind of architecture that unifies neural networks and fuzzy controllers. Some of these approaches have problems when it comes to the interpretation of the learning results. This is especially true, when a pure neural architecture is used.

The NEFCON model presented in this paper has the advantage to be both interpretable as a neural network with fuzzy sets as its weights, and as a fuzzy controller. The learning algorithm based on this model does not result in structural changes, and does not affect the semantics of the underlying fuzzy controller. The learning procedure uses a *fuzzy error measure* to change the fuzzy sets in the system. The fuzzy error describes the current state of a system to be controlled, and controls a reinforment learning algorithm as it is similarly used for neural networks.

2 The NEFCON–Model

The following considerations refer to a technical System S of n state variables $\xi_1 \in X_1, \ldots, \xi_n \in X_n$ and one control variable $\eta \in Y$. Each set $X_i, i = 1, \ldots, n$ is partitioned by p_i fuzzy sets $\mu_1^{(i)}, \ldots, \mu_{p_i}^{(i)}$, and the set Y is partitioned by q fuzzy sets $\nu_1, \ldots, \nu_q$ which are associated with the linguistic terms $A_1^{(i)}, \ldots, A_{p_i}^{(i)}$, and $B_1, \ldots, B_q$ respectively. The knowledge about the output variable η is described by k linguistic control rules $R_1, \ldots, R_k$ [Kruse et al., 1994].

Fig. 1 displays an example of a NEFCON system. The model is consistent with a tree–layer feedforward neural network. The units of the

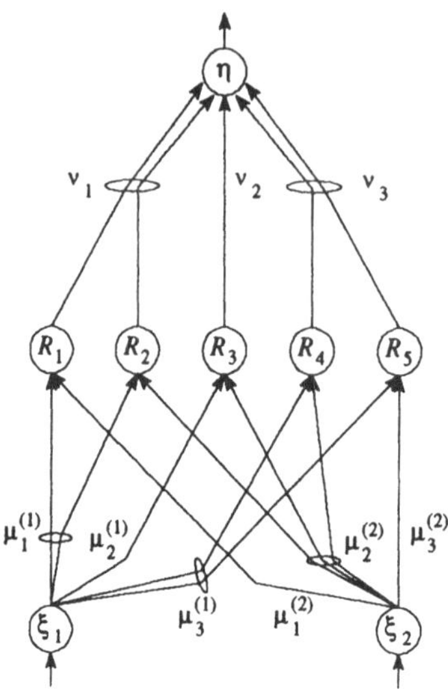

Figure 1: A NEFCON system with two inputs and five rules

input layer just receive and propagate the external input values without further processing. But in case that the input values have to be transformed, the input units have to do it. The units of the hidden layer represent the linguistic rules, and their activations are determined by the matching degrees of their antecedents. The output layer has only one output unit that delivers the crisp value used as a control action applied to the dynamical system S.

The connections between the layers of a NEFCON system are different compared to usual neural networks. There are links which share a common weight. In Fig. 1 e.g. the connections between the input unit ξ_1 to the hidden units R_1 and R_2 carry the same weight $\mu_1^{(1)}$. A learning algorithm has to consider shared weights, and has to apply the same changes to all links that share a common weight.

The weights of the connections are fuzzy sets. The weights between input and hidden layer represent the rule's antecedents, and the weights between hidden and output layer are the fuzzy sets of the rule's conclusions. This way the rule base is encoded by the connections, i.e. by the structure of the NEFCON system.

NEFCON has to be interpreted as a special case of a *three-layer fuzzy perceptron.* To a certain extent, a fuzzy perceptron can be viewed as a "fuzzification" of a normal multi–layer perceptron. It is also used for function approximation, but it has the advantage that its structure can be interpreted in the form of linguistic rules.

Definition 1 *A 3-layer fuzzy perceptron is a 3-layer feedforward neural network* $(U, W, \mathrm{NET}, A, O, \mathrm{ex})$ *with the following specifications:*

(i) $U = U_1 \cup U_2 \cup U_3$ *is a non-empty set of units (neurons), with* $U_i \neq \emptyset$ *and* $U_i \cap U_j = \emptyset$ *for* $i \neq j$, $i \in \{1,2,3\}$. U_1 *is called input layer,* U_2 *rule layer (hidden layer), and* U_3 *output layer.*

(ii) The structure of the network (connections) is defined as $W : U \times U \to \mathcal{F}(\mathbf{R})$, *such that there are only connections* $W(u,v)$ *with* $u \in U_i$, $v \in U_{i+1} (i \in \{1,2\})$.

(iii) A defines an activation function A_u *for each* $u \in U$ *to calculate the activation* a_u

(a) for input and rule units $u \in U_1 \cup U_2$:

$$A_u : \mathbf{R} \to \mathbf{R}, \quad a_u = A_u(\mathrm{net}_u) = \mathrm{net}_u,$$

(b) for output units $u \in U_3$:

$$\begin{aligned} A_u &: \mathcal{F}(\mathbf{R}) \to \mathcal{F}(\mathbf{R}), \\ a_u &= A_u(\mathrm{net}_u) = \mathrm{net}_u. \end{aligned}$$

(iv) O defines for each $u \in U$ *an output function* O_u *to calculate the output* o_u

(a) for input and rule units $u \in U_1 \cup U_2$:

$$O_u : \mathbf{R} \to \mathbf{R}, \quad o_u = O_u(a_u) = a_u,$$

(b) for output units $u \in U_3$:

$$\begin{aligned} O_u &: \mathcal{F}(\mathbf{R}) \to \mathbf{R}, \\ o_u &= O_u(\mathrm{net}_u) = \mathrm{DEFUZZ}_u(\mathrm{net}_u), \end{aligned}$$

where DEFUZZ_u *is a suitable defuzzification function.*

(v) NET *defines for each unit* $u \in U$ *a propagation function* NET_u *to calculate the net input* net_u

(a) for input units $u \in U_1$*:*

$$\mathrm{NET}_u : \mathbf{R} \to \mathbf{R}, \quad \mathrm{net}_u = ex_u,$$

(b) for rule units $u \in U_2$*:*

$$\begin{aligned} \mathrm{NET}_u &: (\mathbf{R} \times \mathcal{F}(\mathbf{R}))^{U_1} \to [0,1], \\ \mathrm{net}_u &= \underset{u' \in U_1}{\top} (W(u',u)(o_{u'})), \end{aligned}$$

where $\top$ *is a t-norm,*

(c) for output units $u \in U_3$*:*

$$\begin{aligned} \mathrm{NET}_u &: ([0,1] \times \mathcal{F}(\mathbf{R}))^{U_2} \to \mathcal{F}(\mathbf{R}), \\ \mathrm{net}_u &: \mathbf{R} \to [0,1], \end{aligned}$$

$$\mathrm{net}_u(x) = \underset{u' \in U_2}{\bot} (\top(o_{u'}, W(u',u)(x))),$$

where $\bot$ *is a t-conorm.*
If the fuzzy sets $W(u',u)$*,* $u' \in U_2$*,* $u \in U_3$*, are monotonic on their support, and* $W^{-1}(u',u)(\tau) = x \in \mathbf{R}$ *such that* $W(u',u)(x) = \tau$ *holds, then the propagation function* net_u *of an output unit* $u \in U_3$ *can alternatively be defined as*

$$\mathrm{net}_u(x) = \begin{cases} 1 & \text{if } x = \dfrac{\sum_{u' \in U_2} o_{u'} \cdot m(o_{u'})}{\sum_{u' \in U_2} o_{u'}} \\ 0 & \text{otherwise} \end{cases}$$

with $m(o_{u'}) = W^{-1}(u',u)(o_{u'})$*. To calculate the output* o_u *in this case*

$$o_u = x, \text{ with } net_u(x) = 1.$$

is used instead of (iv.b).

(vi) $\mathrm{ex} : U_1 \to \mathbf{R}$*, defines for each input unit* $u \in U_1$ *its external input* $\mathrm{ex}(u) = \mathrm{ex}_u$*. For all other units* ex *is not defined.*

The idea of a fuzzy perceptron is the foundation of the NEFCON model [Nauck, 1994, Nauck and Kruse, 1994a]. To create a neural fuzzy controller the fuzzy controller either has to be represented by a fuzzy perceptron, or the fuzzy perceptron has to be interpreted in terms of a fuzzy controller. For that, restrictions on the choice of connections and the determination of the fuzzy weights have to be defined. It is also necessary to define linguistic terms to be associated with the weights, or the connections, respectively.

Definition 2 *Consider a dynamical system S with n variables and one control variable. There are k known linguistic rules describing the control actions applicable to S. A NEFCON system is a fuzzy perceptron with labeled connections adhering to the following constraints.*

(i) $U_1 = \{\xi_1, \ldots, \xi_n\}$, $U_2 = \{R_1, \ldots, R_k\}$, $U_3 = \{\eta\}$.

(ii) Each connection between units $\xi_i \in U_1$ and $R_r \in U_2$ is labelled with a linguistic term $A_{j_r}^{(i)}$ ($j_r \in \{1, \ldots, p_i\}$).

(iii) Each connection between units $R_r \in U_2$ and the output unit η is labelled with a linguistic term B_{j_r} ($j_r \in \{1, \ldots, q\}$).

(iv) Connections comming from the same input unit ξ_i and having identical labels, bear the same weight at all times. These connections are called linked connections. *An analogous condition holds for the connections leading to the output unit η.*

(v) Let $L_{u,v}$ denote the label of the connection between the units $u \in U_1$ and $v \in U_2$. For all $v, v' \in U_2$ holds:

$$((\forall\ u \in U_1)\ L_{u,v} = L_{u,v'}) \Longrightarrow v = v'.$$

This definition allows to interpret a NEFCON system in terms of a fuzzy controller. Condition (iv) makes sure that identical linguistic values of a variable are represented by exactly one fuzzy set. Condition (v) determines that there are no rules with identical antecedents.

The procedures within the neural fuzzy controller correspond to those of a feedforward neural network. The input units represent the crisp input values and propagate them to the rule units. Their propagation function calculates the membership values with respect to the fuzzy sets of the connections between input and hidden layer, and derives the matching degrees of the respective rule antecedents by a t-norm. These values are represented by the activations of the rule units.

The rule units pass their activation values on to the output unit, and its propagation function combines the matching degree with the fuzzy set of the respective link between rule and output unit, calculating an output fuzzy set this way. At last the single rule outputs are accumulated to the overall output in a way that is consistent with the fuzzy control model and the learning algorithm.

The learning procedure needs a crisp output value from each rule. This means there has to be a defuzzification before the accumulating the single results, or the membership functions of the conclusions are monotonous over their support so they can be inverted. NEFCON uses the second alternative (see def. 1).

3 Fuzzy Error Backpropagation – The Learning Algorithm

The goal of the learning algorithm is to adapt the membership functions of the controller, given adequate linguistic control rules but a non-optimal modelling of the fuzzy sets due to a lack of knowledge [Nauck and Kruse, 1992a, Nauck and Kruse, 1992b]. The task of the controller is to drive the system S to an optimal state. But usually a state of the system is also conisidered as *good*, if this optimum is only reached approximately. Therefore it is adequate to describe the goodness of a state by linguistic terms that are represented by fuzzy sets. By this a fuzzy error can be derived that characterizes the performance of the neural fuzzy controller.

There are two cases in the judgement of the system state. In the first case the system S is in a near optimal state, where all variables have near optimal values. In the second state they may have undesired values, which compensate each other, however, in a way that the system is on its way to an optimal state (compensative situation).

Definition 3 *Consider a dynamical system S with n state variables $\xi_1 \in X_1, \ldots, \xi_n \in X_n$, and s known compensative situations. Furthermore there are given n fuzzy sets $\mu_{\mathrm{opt}}^{(i)} : X_i \to [0,1]$, $(i \in \{1, \ldots, n\})$ and s n-ary fuzzy relations $\mu_{\mathrm{komp}}^{(j)} : X_1 \times \ldots \times X_n \to [0,1]$, $(j \in \{1, \ldots, s\})$, describing the optimal values of the state variables, and the compensative combinations of values, respectively. The current state values are given by $(x_1, \ldots, x_n)$. The fuzzy goodness G of the system S is defined as*

$$\begin{aligned} G &: X_1 \times \ldots \times X_n \to [0,1], \\ G(x_1, \ldots, x_n) &= g\left(G_{\mathrm{opt}}(x_1, \ldots, x_n), G_{\mathrm{comp}}(x_1, \ldots, x_n)\right), \end{aligned}$$

where g is a suitable function to combine the measures of goodness with respect to S. G consists of the optimal fuzzy goodness G_{opt}:

$$\begin{aligned} G_{\text{opt}} &: X_1 \times \ldots \times X_n \rightarrow [0,1], \\ G_{\text{opt}}(x_1, \ldots, x_n) &= \top \left\{ \mu_{\text{opt}}^{(1)}(x_1), \ldots, \mu_{\text{opt}}^{(n)}(x_n) \right\}, \end{aligned}$$

and the compensative fuzzy goodness G_{komp}:

$$\begin{aligned} G_{\text{komp}} &: X_1 \times \ldots \times X_n \rightarrow [0,1], \\ G_{\text{komp}}(x_1, \ldots, x_n) &= \top \left\{ \mu_{\text{komp}}^{(1)}(x_1, \ldots, x_n), \ldots, \mu_{\text{komp}}^{(s)}(x_1, \ldots, x_n) \right\}. \end{aligned}$$

The fuzzy error *E of a NEFCON system controlling S is defined as*

$$E : X_1 \times \ldots \times X_n \rightarrow [0,1], \;\; E(x_1, \ldots, x_n) = 1 - G(x_1, \ldots, x_n).$$

Instead of using the definition 3 it is also possible, to describe the fuzzy error by means of linguistic rules. This is done by NEFCON-I. The learning algorithm based on the above defined fuzzy error is called fuzzy error backpropagation. The learning procedure can be compared to backpropagation, but there is no error value depending directly on the output value, so we have a type of reinforcement learning.

The individual part that each rule has in the control value is used to determine the changes by the learning algorithm. This is difficult, if the conclusions are accumulated to a single fuzzy set that has to be defuzzified. Most neural fuzzy approaches that do not use stochastic learning procedures therefore use either special membership functions [Berenji, 1992] or special defuzzification procedures [Berenji and Khedkar, 1992] that are applied before the accumulation.

The approach discussed here uses Tsukamoto's monotonous membership functions for the conclusions which enables a rule to directly supply a crisp value according to its matching degree [Lee, 1990a, Lee, 1990b].

For each rule unit R_r with an activation $o_r > 0$ it has to be decided, if its part t_r it has in the control value has a positive or negative influence, resulting in a "reward" or a "punishment". In addition to the fuzzy error it has to be known, whether the (unknown) optimal control value η_{opt} has to be positive or negative.

The amount of change for the membership function of each rules depends on its *fuzzy rule error:*

$$E_{R_r} = \begin{cases} -o_{R_r} \cdot E & \text{if } \operatorname{sgn}(t_r) = \operatorname{sgn}(\eta_{\text{opt}}) \\ o_{R_r} \cdot E & \text{otherwise.} \end{cases}$$

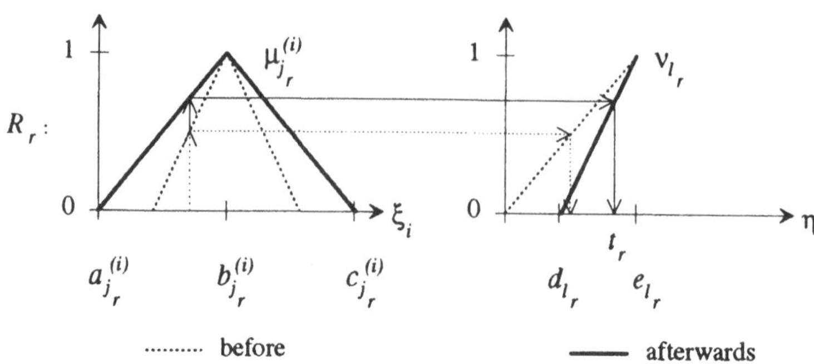

Figure 2: Adapting the membership functions (rewarding)

The changes in the conclusions and the antecedents are defined as follows (see also Fig. 2):

$$\Delta d_{j_r} = \sigma \cdot E_{R_r} \cdot |d_{j_r} - e_{j_r}|$$

where $\sigma > 0$ is a learning rate, and R_r is connected to η by ν_{j_r}.

$$\begin{aligned} \Delta\, a_{j_r}^{(i)} &= -\sigma \cdot E_{R_r} \cdot (b_{j_r}^{(i)} - a_{j_r}^{(i)}), \\ \Delta\, c_{j_r}^{(i)} &= \sigma \cdot E_{R_r} \cdot (c_{j_r}^{(i)} - b_{j_r}^{(i)}), \end{aligned}$$

where ξ_i is connected to R_r by $\mu_{j_r}^{(i)}$. Shared weights are changed repeatedly. In [Nauck and Kruse, 1993, Nauck et al., 1994] an extension of the learning algorithm is presented, that enables the model to learn fuzzy rules, too.

4 NEFCON–I – An Implementation

NEFCON–I is a graphical simulation environment for NEural Fuzzy CONtrollers developed at the Dept. of Computer Science, TU Braunschweig (see Fig. 3). It lets a user define a neural fuzzy controller and train it [Nauck and Kruse, 1994b]. The user defines the linguistic variables and an initial partioning with membership functions. He can decide to enter the rule base or to let NEFCON–I learn the rules. After the fuzzy error is defined in terms of fuzzy rules, NEFCON–I can be trained

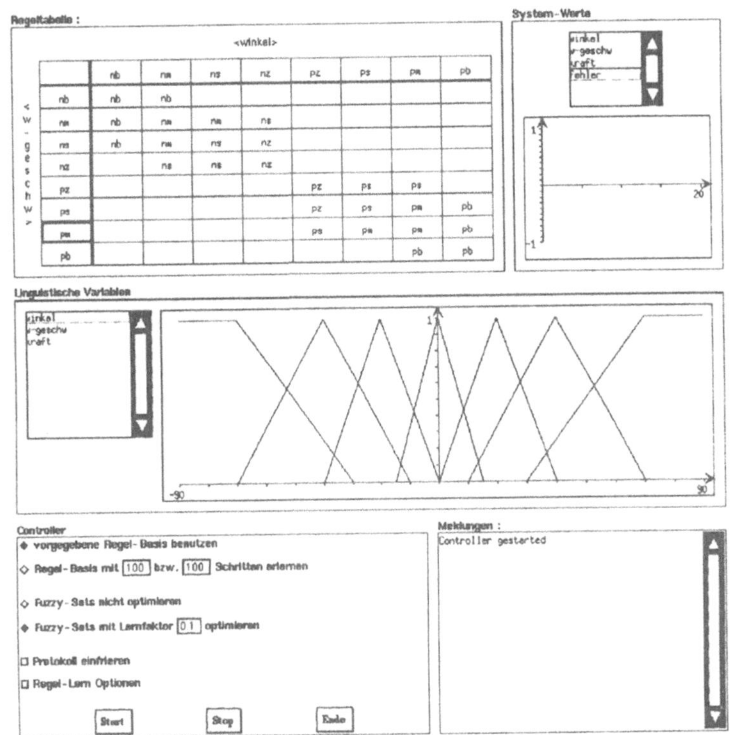

Figure 3: The user surface of NEFCON–I

by exchanging data with a concurrently running program that simulates a dynamical system.

To test the model a simplified version of the inverted pendulum was used, considering only angle and angular velocity. Fig. 4 shows the partioning of the variable "angle" before and after the learning process. The controller was not able to balance the pendulum at the beginning, but learned to do it in about 2000 cycles (2 min. on a SUN workstation). The changes in the control surface of the neural fuzzy controller are shown in Fig. 5.

NEFCON–I is freely available for scientific purposes by anonymous ftp. If you have internet access you can download the source code or one

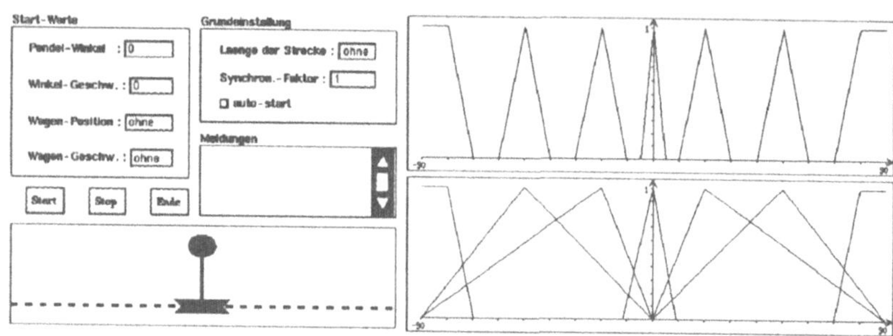

Figure 4: Pendulum and fuzzy sets for *angle* before and after learning

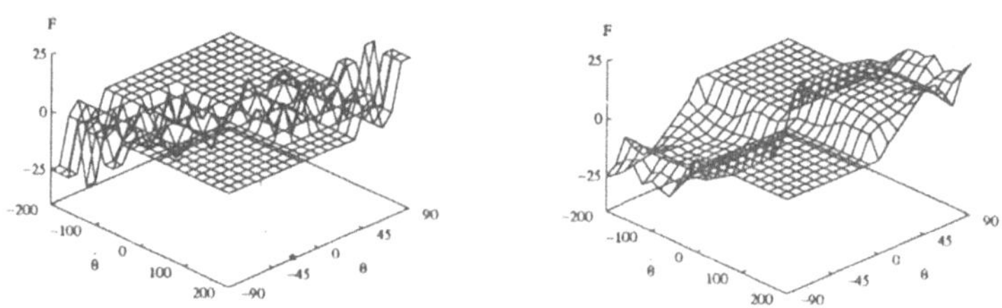

Figure 5: The control surface before and after learning

of the binaries for special platforms. To compile NEFCON–I you need a UNIX workstation running X–Window, and you need the graphical software library InterViews 3.1. To run one of the binaries just a Unix workstation under X–Window is needed.

To obtain NEFCON–I you need to connect to ftp.ibr.cs.tu-bs.de (134.169.34.15) by anonymous ftp, and change to the directory /pub/local/nefcon. There you will find the necessary instructions what to download, and how to proceed.

References

H. R. Berenji (1992). *A Reinforcement Learning-Based Architecture for Fuzzy Logic Control*. Int. J. Approximate Reasoning, 6:267 –292.

H. R. Berenji and P. Khedkar (1992). *Learning and Tuning Fuzzy Logic Controllers Through Reinforcements*. IEEE Trans. Neural Networks, 3:724–740.

R. Kruse, J. Gebhardt and F. Klawonn (1994). *Foundations of Fuzzy Sets*. Wiley, Chichester.

C. C. Lee (1990a). *Fuzzy Logic in Control Systems: Fuzzy Logic Controller, Part I*. IEEE Trans. Systems, Man & Cybernetics, 20:404–418.

C. C. Lee (1990b). *Fuzzy Logic in Control Systems: Fuzzy Logic Controller, Part II*. IEEE Trans. Systems, Man & Cybernetics, 20:419–435.

D. Nauck (1994). *Modellierung Neuronaler Fuzzy–Regler (Modeling Neural Fuzzy Controllers [in German])*. PhD thesis, Technische Universität Braunschweig.

D. Nauck, F. Klawonn and R. Kruse (1994). *Neuronale Netze und Fuzzy–Systeme (Neural Networks and Fuzzy Systems [in German])*. Vieweg, Wiesbaden.

D. Nauck and R. Kruse (1992a). *Interpreting Changes in the Fuzzy Sets of a Self-Adaptive Neural Fuzzy Controller*. In *Proc. Second Int. Workshop on Industrial Applications of Fuzzy Control and Intelligent Systems (IFIS'92)*, pages 146–152, College Station, Texas.

D. Nauck and R. Kruse (1992b). *A Neural Fuzzy Controller Learning by Fuzzy Error Propagation*. In *Proc. Workshop of North American Fuzzy Information Processing Society (NAFIPS92)*, pages 388–397, Puerto Vallarta.

D. Nauck and R. Kruse (1993). *A Fuzzy Neural Network Learning Fuzzy Control Rules and Membership Functions by Fuzzy Error Backpropagation*. In *Proc. IEEE Int. Conf. on Neural Networks 1993*, pages 1022–1027, San Francisco.

D. Nauck and R. Kruse (1994a). *A fuzzy perceptron as a generic model of multilayer fuzzy neural networks*. In *Proc. Fuzzy–Systeme'94*, Munich.

D. Nauck and R. Kruse (1994b). *NEFCON–I: An X–Window based Simulator for Neural Fuzzy Controllers*. In *Proc. IEEE Int. Conf. Neural Networks 1994 at IEEE WCCI'94*, Orlando.

4

Fuzzy Systems in AI

4.1. Fuzzy Systems in AI: An Overview

Christian Freksa

Abstract

This paper reviews motivations for introducing fuzzy sets and fuzzy logic to knowledge representation in artificial intelligence. First we consider some areas of successful application of conventional approaches to system analysis. We then discuss limitations of these approaches and the reasons behind these limitations.

We introduce different levels of representation for complex systems and discuss issues of granularity and fuzziness in connection with these representation levels. We make a distinction between decomposable and integrated complex systems and discuss the relevance of this distinction for knowledge representation and reasoning. We also distinguish fuzzy relations between quantities of different granularity within one domain from fuzzy relations between two different domains and discuss the need of considering both in artificial intelligence.

We distinguish methods for describing natural, artificial, and abstract systems and contrast the modeling of system function with the modeling of system behavior in connection with the representation of fuzziness. The paper briefly discusses recent criticism of the fuzzy system approach and concludes with a prospect on soft computing in AI.

1 Why do we Need Fuzzy Sets and Fuzzy Logic in AI?

The notion of a fuzzy set [Zadeh 1965] and the development of *fuzzy set theory* and *fuzzy logic* were motivated by the severe difficulties to adequately characterize complex systems by conventional approaches of system analysis. "Adequate" means, for example, that insignificant variations on the component level of a system should not add up to significant changes on the system level. This criterion is an absolute requirement for understanding complex systems in terms of their components.

Conventional approaches represent complex systems in a reductionist manner by specifying well-defined components and their individual interactions. We will investigate the question why these approaches are of limited use in artificial intelligence and cognitive science.

The success of conventional approaches in well-defined systems

Until fairly recently, most researchers in artificial intelligence believed, AI systems could benefit from the same virtues that made much of modern science and technology a success: precise measurements, complete knowledge of the domain, and rigorous tools for dealing with them. Before discussing this belief I will briefly investigate under which conditions dealing with precise, crisp, and complete measurements has been successful.

A fundamental prerequisite for succeeding with high-precision characterizations of complex systems is the availability of basic entities and relations suitable for capturing everything that is relevant in the system under consideration. Under this condition, we can decompose complex systems into less complex subsystems and/or basic components and we can explain their role in terms of the basic components and their interactions.

Obviously, we can correctly describe complex systems in terms of their components when (1) the components conform with their definitions, (2) the interactions conform with their specifications, and (3) no other interactions interfere. These conditions certainly hold in *abstract systems* that are specified according to the three conditions given above. Examples are complex games defined by simple rules like the board games chess and go.

The three conditions also can be assumed to hold in *technical systems* whose components can be studied in isolation and whose interactions can be restricted to controlled local exchanges. Examples are closed chemical, mechanical, and electronic systems. Complex computers and computer networks constitute excellent examples that the bottom-up approach to characterizing systems on the basis of crisp notions can be highly successful.

However, the more complex these systems get, the more obvious becomes the need for developing high-level views and languages to better capture the essential aspects including actions at the relevant system level. Thus, the granularity of a description can make a big difference even if fine and coarse descriptions of crisp systems can be considered equivalent, in a formal sense.

The problem with ill-defined systems

The great success in representing complex systems in terms of their simple components that could be achieved in closed *technical systems* may be responsible for a blind belief that the same approach could be applied to other complex systems equally well. Let us consider a class of non-technical complex systems whose representation in terms of components has caused great difficulties: systems of economics or climate systems may serve as examples. Why are such systems different?

In case of an economic system, the three above requirements seem fulfilled, at first glance: (1) monetary units define the economic value of any item in basic terms, (2) mathematical rules precisely derive the effects of economic transactions, and (3) no other interactions besides transactions determine the monetary values on the component level. In fact, the strict obedience of the mathematical laws by economic transactions make it possible to precisely analyze in quantitative terms why someone became rich, became poor, went bankrupt, etc.

The problem is, that it is not very interesting to have models that only can be used for *post hoc* analysis. In the case of climate systems even the post hoc analysis does not carry very far. We would like to build models to make predictions! Why is it possible to make predictions on the component level for board games but not for economic systems?

The role of system complexity

The fact is that practically speaking it is not even possible for board games like chess or go to make reliable predictions on the component level of description. The complexity of the chess game consisting of only 32 pieces on an eight by eight checkers board already is too big for analyzing all legal moves. Nevertheless, it is possible to build useful higher-level descriptions of chess constellations from the basic entities and use these descriptions to generate reasonable predictions.

In systems of economy, in contrast, this approach has not proven practicable; condition (3) is severely violated: individual transactions in economy systems can not be predicted on the basis of local transactions; they are determined by complex global patterns and their dynamics involving psychological and other factors which cannot be captured in terms of the elementary configurations.

A similar situation is given in the case of climate systems and weather prediction: our knowledge about the preconditions in terms of fundamental facts will never be complete enough and our knowledge about the physical laws on the local level does not suffice for making useful predictions about global or local weather conditions. Thus, system complexity is only one aspect that must be considered; the availability of knowledge and the structure of that knowledge also play an import role for the representation we can use.

How does fuzziness come into the picture?

If we compare systems that are well-defined (from the bottom up) with systems that we know primarily from their global properties, we find different mapping relations between the low-level and the high-level notions. The high-level notions built up from low-level primitives typically are found in a *crisp* relation to the synthesized complex notions while the low-level notions postulated from the high-level concepts are found in a *fuzzy* relation.

For example, a taxonomy of plants and/or animals based on low-level primitives will yield a crisp classification of creatures as found in biology textbooks. On the other hand, the identification of low-level features for the definition of the high-level everyday notion "living animal" yields fuzzy relations between the high-level notion and the low-level features, as it is impossible to precisely capture the everyday notion universally by composition of low-level features. In this sense, "living animal" is a fuzzy concept, when related to low-level primitives – while on the high level on which we typically use the notion it would not be considered fuzzy at all.

2 The level of representation

Systems about whose properties we learn from global behavior can be described meaningfully on a global level, for example in the case of an economic system we might know "when the interest rate goes up, the money flow decreases". Such a rule implicitly can take into account complex interaction regularities which cannot be captured on the component level. Although the total money flow results from individual monetary transactions, the rule does not hold for each individual transaction. Therefore it is difficult to give precise definitions of the global notions in terms of local transactions; it is much easier to identify the global effect as the net effect of local transactions and to control economy on a global level (e.g. by manipulating the interest rate) than through local transactions.

In the case of weather prediction we encounter a similar situation: on a coarse level (which may be relevant for agriculture, for example), predictions may be quite reliable, while they may be useless on the level of description on which local measurements of weather indicators (like rainfall per square meter) are made.

As a consequence, by representing complex systems not on the level of the most primitive notions but on a coarser level, these systems may become tractable. Numerous complex interactions on the local level simply can be ignored! Lotfi Zadeh recognized the importance of abstraction from low-level properties at an early stage of the artificial intelligence enterprise. He characterized intelligent systems by their ability to summarize complex descriptions by abstracting from details. In the case of everyday non-synthetic systems this requires taking into account the fuzzy relations between the high-level and the low-level features.

Integrated complex systems

There is a class of complex systems in which fuzzy relations play a particularly important role. This class consists of systems whose components cannot be studied in isolation or whose components cannot be studied in all relevant conditions. Most *natural complex systems* belong to this class. We can observe global behavior under varied condition patterns; from these we infer local influences.

Examples are biological systems which we describe in terms of presumed local functions and observed effects. Neither the description of the global effects nor the description of the local functions are suitable to capture all possible situations and to crisply represent the system. The reason is, that we are bound to use concepts which have a meaning outside of these systems (like "living animal") since elementary local components are not available and there is no way to guarantee that these concepts precisely match the components of the described system.

Fuzzy relations between different domains

Fuzzy pattern recognition, fuzzy control, and most other successful applications of fuzzy set theory have focused on the fuzzy relation between the fine and coarse levels of representation of the artifacts involved. But from the inception of fuzzy set theory, its inventor Zadeh also suggested to represent fuzzy relations between *real* entities like physical objects and *mental* entities like concepts which are manifested in natural language expressions. In artificial intelligence, this type of relations is of particular interest.

Classical artificial intelligence had been focusing its attention on the representation structures within the medium computer. Little attention had been paid to ontological questions and the actual representation problem, i.e. the mapping between what is represented and what is representing [c.f. Palmer 1978] and to the problem of building representation structures from existing knowledge. Many AI-researchers did not consider this issue a problem; they believed, concepts could be used like nuts and bolts to screw together intelligent systems and the role of each concept would be clearly defined.

But when AI-systems grew up and moved outside their purely synthetic laboratory environments it became evident that there was a serious matching problem between natural concepts derived from the use and the behavior of systems and artificial concepts synthesized from low-level components. It became clear that we could not simply view natural concepts as imperfect entities whose objectives would be much better served by artificial substitutes.

Instead, expert system research and research in cognitive science investigated structures and properties of human knowledge in order to exploit its potential and to understand more of its function. In this process, fuzzy sets have played an important role in characterizing the relation between human concepts and natural or artificial entities and in mimicking their interactions (c.f. [Zimmermann 1992], [Dubois et al. 1993], [Kruse et al. 1994]). In this way, many of the properties of human thought and natural language, specifically with regard to their modification and their combination, could be simulated.

3 Natural systems vs. artifacts

As argued in the previous sections, the development of fuzzy set theory and fuzzy logic were influenced by properties of natural human concepts and their relation to entities in the real world. Specifically, Zadeh suggested to use fuzzy sets to represent notions like *tall* and *beautiful* in natural language phrases like *The tall dwarf is more beautiful than the small giant*. However, the present success in the application of fuzzy set theory is not so much in artificial intelligence – e.g. in the representation of natural language expressions or human concepts – but rather in control engineering, e.g. due to the improvement of purely artificial systems like household appliances, photo equipment, trains, and helicopters [Munakata, Jani 1994]. Why have fuzzy sets not caught on in artificial intelligence to the same degree as in control engineering?

The utility of fuzzy sets in artificial systems is not only due to reduced complexity as a consequence of the coarser representation of systems, but also due to properties of human concepts which support the engineering process of these systems. Fuzzy sets serve as a knowledge transfer vehicle, as a way of getting the judgment of engineers into complex systems. They are particularly suited for this task since they allow for reasonable representation relations even if the representation system is not yet completely understood. In this sense, human concepts are involved in the synthetic products of the engineers. But does this mean that fuzzy sets represent human concepts?

The characteristic function which defines a fuzzy set *characterizes* the relationship between real world entities and specific concepts (or labels typically associated with concepts); however, it does not model or explain how this fuzzy relationship comes about. For certain domains or types of tasks, a characterization of the fuzzy relationship between a concept and given instances in 'reality' is sufficient (for some tasks even this relationship is not required) - but there is an important class of problems which requires going beyond the characteristic function.

Shallow vs. deep modeling

I will argue that classical fuzzy sets represent human concepts on a rather shallow level, on the level of denoting entities within a well-defined framework, a framework in which the relevant dimensions are known, but precise values within these dimensions are missing. Human concepts, however, are not merely underdetermined physical values. They form a strong system on their own which have a meaning and make sense independent of real-world instances even though during the knowledge acquisition process they may have been derived from such instances.

In Figure 1, I present a classification of different epistemic qualities of knowledge associated with methods appropriate for processing them. On the basis of this classification I will suggest ways of extending the fuzzy set philosophy towards representing human concepts on a deeper level. The goal is to adapt notions developed in fuzzy set theory and fuzzy logic to make them better suitable for processing human knowledge and knowledge representation in artificial intelligence.

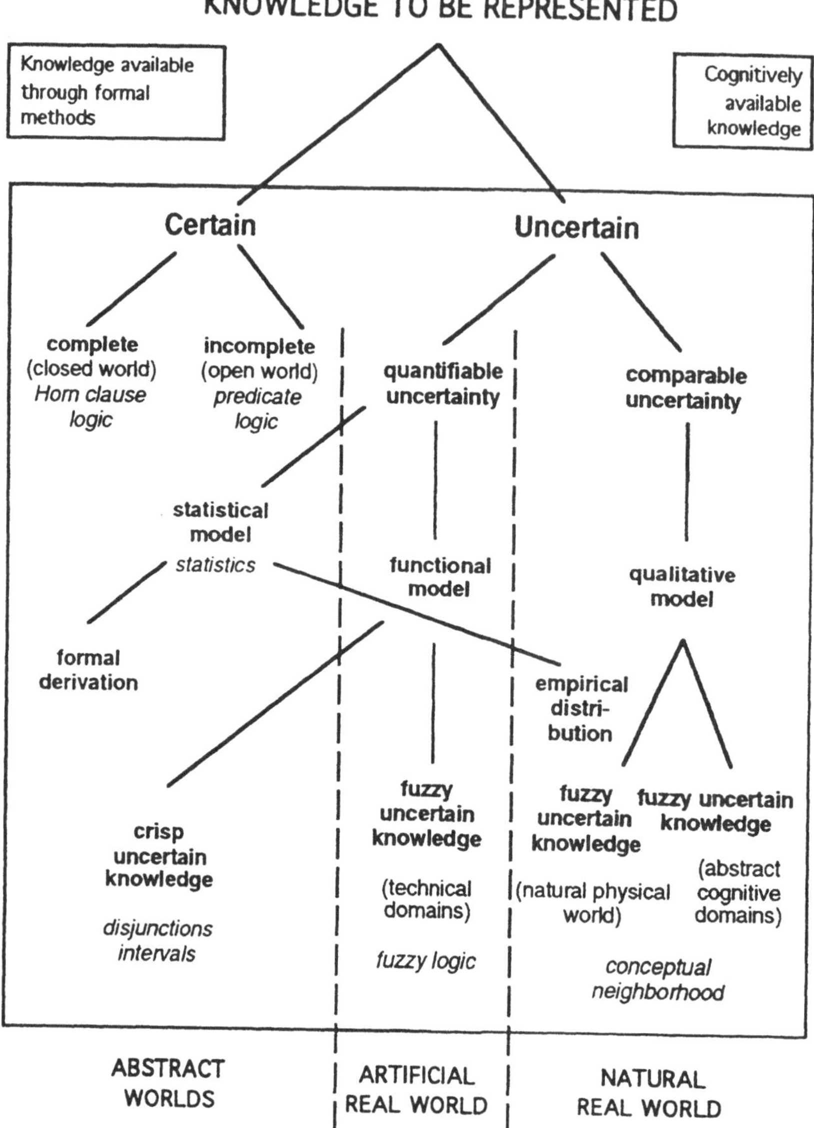

Figure 1: Classification of knowledge types on the basis of accessibility criteria

A classification of knowledge types

The classification of knowledge types presented in Figure 1 distinguishes different types of situations in which knowledge can be acquired. I will first present the distinctions made in this classification and subsequently I will associate types of approaches which appear suitable for dealing with the respective situations and knowledge.

The top-level distinction in this classification is made between *certain* knowledge and *uncertain* knowledge. Certain knowledge is only available in *abstract* domains where facts and rules can be postulated to be true. Knowledge obtained from *concrete* domains are subjected to uncertainty, due to limitations in modeling and knowledge acquisition. Certain knowledge may be further classified into *complete* knowledge in which the non-existence of a true fact is equivalent to the existence of the negated fact *(closed world assumption)* and *incomplete knowledge* in which this assumption is not generally valid.

Uncertain knowledge can be further classified into knowledge with *quantifiable* uncertainty, i.e. knowledge with which we can associate an absolute degree of uncertainty, and *comparable* uncertainty, i.e. knowledge that we merely can rank according to relative degrees of uncertainty.

Uncertainty may be due to statistical effects, due to incomplete knowledge, or due to fuzziness; accordingly uncertainty can be quantified on the basis of statistical models or on the basis of functional models. Statistical models can be classified according to the source of the probability distribution: are the probability distributions derived from a *formal* model on the basis of assumptions about the processes involved (e.g. each face of a die is equally likely to be thrown) or from empirical observations.

In the class of functional models of uncertainty we can distinguish between those yielding *crisp* sets of possible values and those yielding *fuzzy* sets. The crisp sets may be either intervals (in the case of continuous domains) or disjunctions (in the case of discrete domains). Fuzzy sets can be obtained when the functional dependencies are clearly enough defined to allow for a quantitative characterization of the elasticity of constraints. This is the case in *synthetic* domains in which fuzzy membership values can be derived from the system specifications. For an interesting illustration of the advantages of studying synthetic systems rather than natural systems, see Braitenberg's experiments in synthetic psychology [Braitenberg 1984].

In the domain of natural language or other natural domains, quantitative membership values generally are not available; but frequently membership values can be compared like in "I would rather say 'John is tall' than 'John is very tall'". Such comparisons lead to qualitative models of uncertainty. We can further distinguish between fuzzy uncertain knowledge about the natural physical world where we may have empirical sensory evidence for comparative uncertainty judgments and fuzzy uncertain knowledge about abstract cognitive domains where it is not possible to identify objective correlates to the comparative judgments.

In summary, this classification yields three different major types of knowledge processing situations: 1) The situation of abstract domains in which we work on the basis of formal specifications; here we either deal with complete certainty or with well-controlled uncertainty. 2) Real-world situations which are well-controlled in the sense that we know all relevant factors and the possible extreme situations; this is the case in many technical domains whose function *in principle* could be completely described on the basic level by exhaustive analysis; such artificial domains are ideal for the application of classical fuzzy logic. 3) In artificial intelligence, we are frequently confronted with situations in the natural world in which the available knowledge does not justify a quantitative fuzzy set approach; in particular, we must deal with open worlds in which representations in terms of numerical membership functions do not make sense; instead, qualitative knowledge may be available. For this type of situation, more adequate representational approaches are needed.

I would like to suggest that fuzzy reasoning may become as successful as fuzzy control and will be useful for natural language processing and other AI applications when it is integrated with deep conceptual representations of the domain and of the language describing the domain. In particular, the 'horizontal' dimension connecting cooperating and competing concepts must be exploited in addition to the 'vertical' dimension which connects concepts with their definitions. This horizontal dimension is required for determining which concepts are to be used in the first place. Depending on this selection, fuzziness may or may not play a role for the given task.

A possibility for extending the scope of fuzzy reasoning on the basis of the notion of *conceptual neighborhood* is currently being explored by the author. The notion of conceptual neighborhood was introduced in the context of qualitative temporal and spatial reasoning [Freksa 1992]. It addresses the logical structure of the represented domain under the operations that can take place in the domain. Specifically, two relations are conceptual neighbors when there is an operation in the represented domain which can transform one of the two relations into the other. Conceptual neighborhood can be used to represent horizontal relationships between concepts and to form coarse concepts from fine concepts and vice versa.

4 Criticism of the fuzzy systems approach

In a prize-winning article, Charles Elkan expressed his surprise about the success of fuzzy logic [Elkan 1993]. Specifically, Elkan attempts to show formally that fuzzy logic collapses to classical two-valued logic and he argues that it is not adequate for reasoning about uncertain evidence in expert systems from an empirical perspective. Elkan's overall judgment is that fuzzy logic is fundamentally wrong and will cause serious problems in more challenging applications.

Like Zeno's famous ancient paradoxes [c.f. Vlastos 1967], Elkan's paradox appears rather convincing at first glance, when the reader submits himself to the formal framework used by the author. However, a more careful analysis of Elkan's argument reveals that – like Zeno – he presents a restricted view. He does this by forcing a notion of equivalence valid for two-valued logic on the analysis of fuzzy logic. In this way, Elkan addresses only special cases in which two-valued and fuzzy logic in fact are equivalent [cf. Shastri, to appear]. It is easy to provide numerical counterexamples to Elkan's equivalence assertion using standard notions of fuzzy sets for intermediate truth values.

Limitations of formal analysis

I will take the occasion of this attack on fuzzy systems on the basis of purely formal arguments to draw your attention to an important issue of knowledge representation systems which *in principle* cannot be resolved by formal analysis. Formal analysis helps us to understand systems which are entirely formal. However, in representing knowledge about the real world, one part of the system is the body of knowledge to be represented, another part is the representing formal structure, and a third part establishes the relations between the body of knowledge and the formal structure [c.f. Palmer 1978].

Only the second part, the formal structure, can be rigorously analyzed formally. The first part, the body of knowledge is not accessible with formal tools directly; human perception and/or intuition present the knowledge to be represented by the formalism. The representation of the knowledge can be only as good as our understanding of the structure of the knowledge itself! For example, if we take a natural language statement like "John is tall" merely as a different way of writing the predicate logic statement *tall (John) is true*, then we will never be able to reach aspects of the original statement which are not coded in the predicate logic 'equivalent'. We easily can become victims of the same kind of fallacy Charles Elkan was subjected to when he viewed the world of fuzzy logic through the glasses of two-valued logic.

In reasoning about the real world, making formally sound inferences is only one aspect. Equally important – but much more difficult than widely believed – is to adequately formalize real world knowledge in the first place. The "paradoxical success" of the fuzzy logic approach in restricted domains may be considered as an indication that clearer perception or sharper intuition about the relation between the domain knowledge and the domain states have been involved in the knowledge formalization process. Of course, having found a new representation structure, we must develop appropriate reasoning methods to go along with.

It may well be impossible to find methods which will both fit the more adequate representation structure for real world phenomena and satisfy the classical criteria of formal analysis such as logical equivalence. Nevertheless, the resulting inferences may be more useful than formally correct inferences on the basis of less adequate knowledge structures. The notion of representational adequacy is not yet sufficiently understood.

5 Soft computing

Fuzziness is one of several aspects of our knowledge about the real world which must be taken into account in knowledge representation and processing. In general, we must deal with imprecision, uncertainty, and partial truth. The human mind can be viewed as a working realization of a system which rather successfully deals with all of these aspects simultaneously. In contrast to conventional (hard) computing approaches, systems that are tolerant of these aspects of everyday knowledge are united by the label *soft computing*.

The guiding principle of soft computing is: Exploit the tolerance for imprecision, uncertainty, and partial truth to achieve tractability, robustness, and low cost solutions [Zadeh 1994]. The basic ideas underlying soft computing have links to many early influences of fuzzy set theory, including Zadeh's original publication on fuzzy sets [Zadeh 1965], his paper on the analysis of complex systems and decision processes [Zadeh 1973], and his paper on possibility theory and soft data analysis [Zadeh 1981].

Besides fuzzy logic, probabilistic approaches for reasoning under uncertainty and related models for belief maintenance and revision play an important role. In artificial intelligence, Pearl's probabilistic reasoning in Bayesian networks, Nilsson's probabilistic logic, the certainty factor model used in the MYCIN expert system for medical diagnosis, Dempster-Shafer's theory of evidence have attracted much attention (c.f. [López de Mántaras 1990], [Kruse et al. 1991]). In the mid 1980s, neural network theory also joined into the soft computing effort.

Combining different approaches to soft computing

It has become evident during the past ten or twenty years, that no single approach to the study of cognitive or intelligent processes will succeed in understanding the interactions of cognitive agents with complex environments and no single approach to representing complex knowledge will fulfill all our requirements. Successful AI approaches must take into account effectiveness, efficiency, timeliness, robustness, adequacy, and cost of the solutions. Classical requirements like provability of correctness and completeness of the solution can be expected as little from computer systems reasoning about complex situations as from humans in the same situation.

After an era of increasing specialization in almost all areas of research and technology, we have now entered an era in which the interaction of approaches is of particular importance and concern. This is true for numerous areas, but interdisciplinary efforts like cognitive science and multi-approach efforts like the Berkeley Initiative in Soft Computing (BISC) might serve as examples. Such efforts require a considerable amount of re-orientation, as we have to recognize that the former competitors must become partners.

Although fuzzy set theory and fuzzy logic have faced strong opposition from conventionally oriented theoreticians in artificial intelligence and logic during the past 30 years, the rapidly growing number of successful applications developed mainly in Japan have shifted the focus of interest from local formalistic concerns to global system considerations. As in the case of the Fifth Generation Computing Project in the early 80s it required the Japanese challenge before European and American opposition was matched by a growing interest in the industry and an increased willingness by theoreticians to understand the principles of soft computing.

In Europe, we now find an increasing interest in the theory and applications of soft computing techniques in artificial intelligence. This is evident from the growing number of fuzzy logic workshops and soft computing contributions to artificial intelligence conferences, from the establishment of special interest groups in fuzzy logic and soft computing (e.g. within the German computer science society) and from the growing number of tutorials offered both by the industry and by academic institutions.

Three papers on specific topics of fuzzy reasoning

In the remainder of this chapter you find three articles dealing with soft computing for artificial intelligence.

The article by Sascha Dierkes, Bernd Reusch, and Karl-Heinz Temme presents a tool for supporting the representation of fuzzy knowledge and for fuzzy reasoning in an experts system shell.

The article by Jörg Gebhardt uses the possibilistic interpretation of fuzzy sets in the context of model-based reasoning. The approach described in the paper allows for evidential reasoning in multidimensional numerical hypothesis spaces under imprecision and uncertainty.

Jochen Heinsohn presents a language to extend the taxonomic knowledge representation approach of terminological logics by a probabilistic knowledge representation component. In this way uncertain knowledge can be included in the reasoning process.

Acknowledgments

I thank Jörg Gebhardt, Jochen Heinsohn, and Ramon López de Mántaras for valuable comments on an earlier version of this paper.

References

[Braitenberg 1984] V. Braitenberg *Vehicles*. MIT-Press, Cambridge 1984.

[Dubois et al. 1993] D. Dubois, H. Prade, R.R. Yager (eds.) *Readings in Fuzzy sets for intelligent systems*. Morgan Kaufmann Publishers, San Mateo 1993.

[Elkan 1993] C. Elkan: The paradoxical success of fuzzy logic. *Proc. AAAI-93*. 698-703, AAAI Press/MIT Press, Menlo Park 1993.

[Freksa 1992] C. Freksa: Temporal reasoning based on semi-intervals, *Artificial Intelligence* 54 (1992) 199-227.

[Kruse et al. 1991] R. Kruse, E. Schwecke, J. Heinsohn: *Uncertainty and vagueness in knowledge based systems: numerical methods*. Series Artificial Intelligence, Springer, Heidelberg 1991.

[Kruse et al. 1994] R. Kruse, J. Gebhardt, F. Klawonn: *Foundations of Fuzzy Systems*. John Wiley and Sons, Chichester, 1994.

[López de Mántaras 1990] R. López de Mántaras: *Approximate reasoning models*. Ellis Horwood, Chichester 1990.

[Munakata, Jani 1994] T. Munakata, Y. Jani: Fuzzy Systems: An overview. *Communications of the ACM* vol. 37, 3, 69-76, 1994.

[Palmer 1978] S.E. Palmer: Fundamental aspects of cognitive representation. In: Rosch, E., Lloyd, B. (eds.), *Cognition and categorization*, Erlbaum, Hillsdale 1978.

[Shastri, to appear] L. Shastri (ed.): Symposium on Fuzzy Logic, *IEEE Expert* (to appear).

[Vlastos 1967] G. Vlastos: Zeno of Elea. In: P. Edwards (ed.) *The Encyclopedia of Philosophy* vol. 8, 369-379. New York: Macmillan 1967.

[Zadeh 1965] L.A. Zadeh: Fuzzy sets. *Information and Control* 8, 338-353, 1965.

[Zadeh 1973] L.A. Zadeh: Outline of a new approach to the analysis of complex systems and decision processes. *IEEE Trans. SMC* 3, 1, 28-44, 1973.

[Zadeh 1981] L.A. Zadeh: Possibility theory and soft data analysis. *Mathematical Frontiers of the Social and Policy Sciences*, L. Cobb and R.M. Thrall (eds.), 69-129. Westview Press, Boulder 1981.

[Zadeh 1994] L.A. Zadeh: Fuzzy logic, neural networks, and soft computing. *Communications of the ACM* vol. 37, 3, 77-84, 1994.

[Zimmermann 1992] H.-J. Zimmermann. *Fuzzy set theory and its applications*. Kluwer, Boston 1992.

4.2. Possibilistic Reasoning in Multidimensional Hypotheses Spaces

Jörg Gebhardt

Abstract

This paper introduces a relational approach to possibilistic reasoning in knowledge-based systems. We consider a *possibilistic focusing system*, where qualitative knowledge on existing dependencies among attributes is represented with the aid of a hypergraph, and uncertainty about these relationships in terms of possibility distributions. Besides an outline of some basic concepts, we discuss an application and compare the framework with Bayesian networks.

1 Introduction

Restricting ourselves to pure numerical environments, this paper introduces a framework for possibilistic reasoning. In particular, section 2 proposes a relational setting for imprecise reasoning in a multidimensional space of attributes. Considering a universe of discourse $\mathfrak{U} = (\Omega_i)_{i \in N}$, where Ω_i denotes the domain of the i–th attribute, and $\Omega \stackrel{\text{Df}}{=} \bigtimes_{i \in N} \Omega_i$ the induced product space, we suppose that *general knowledge* R about relationships between attribute values as well as *evidential knowledge* E about a specific object state $\omega_0 \in \Omega$ under consideration is characterized in terms of relations $R \subseteq \Omega$ and $E \subseteq \Omega$. Then we introduce the concept of a *focusing system*, which specifies qualitative dependencies between attributes by means of a hypergraph $(N, \mathfrak{M})$, whereas quantitative dependencies are expressed by a rule system $\mathcal{R}(\mathfrak{U}, \mathfrak{M})$, consisting of relations R^M, referred to the attributes identified by all indices contained in $M \in \mathfrak{M}$.
Incorporating uncertainty aspects and therefore generalizing from imprecise to possibilistic knowledge, section 3 introduces the concept of a *possibilistic focusing system* and its semantic background.

This work has been partially funded by CEC–ESPRIT III Basic Research Project 6156 (DRUMS II)

Finally, section 4 is a short discussion of possibilistic vs. probabilistic reasoning in causal networks. In this connection we outline an example, where reasoning in a possibilistic focusing system has turned out to be a promising alternative of its probabilistic counterpart.

2 A Relational Setting for Imprecise Reasoning

Let $\mathcal{A} = \{a_1, \ldots, a_n\}$ be a set of attributes that characterize an object under consideration and $\Omega_i = Dom(a_i)$, $i = 1, \ldots, n$, their attached (finite) domains. Stating the *closed world assumption*, the current state of this object is expected to be specifiable in terms of a tuple $(\omega_0{}^{(1)}, \ldots, \omega_0{}^{(n)}) = \omega_0$, with $\omega_0{}^{(i)} \in \Omega_i$ and $\omega_0 \in \Omega \stackrel{\text{Df}}{=} \bigtimes_{i=1}^{n} \Omega_i$.
Independent from what kind of reasoning problem we intend to investigate, it is appropriate in our setting to presuppose that *general knowledge* about relationships between the chosen attributes is formalized by means of a *relation* $R \subseteq \Omega$, whereas *evidential knowledge* (to be interpreted as additional information on the specific object state ω_0) is specified by another relation $E \subseteq \Omega$.

In standard cases of practical interest, R and E will of course not directly be available, but rather occur as the result from combining smaller pieces of knowledge R_j, $j = 1, \ldots, r$, that refer to *quantitative dependencies* between the possible values of subsets $\mathcal{A}_j \subseteq \mathcal{A}$ of attributes, where $\mathcal{A}_j$ reflects an existing *qualitative dependency* with respect to the attributes contained in $\mathcal{A}_j$.

To support a nice handling of R_j we choose the following formal environment:

Let $\mathcal{A}' \subseteq \mathcal{A}$ be a set of selected attributes and

$$I(\mathcal{A}') \stackrel{\text{Df}}{=} \{i \in \{1, \ldots, n\} \mid a_i \in \mathcal{A}'\}$$

its *identifying index set*. As abbreviation we introduce the special index set $N = \{1, \ldots, n\}$.

An arbitrary set $\mathfrak{M} = \{I_1, \ldots, I_r\}$ of index sets is called a *modularization* of N, iff

(a) $(\forall I \in \mathfrak{M})\,(\emptyset \neq I \subseteq N)$,

(b) $\bigcup_{I \in \mathfrak{M}} I = \{1, \ldots, n\}$,

(c) $(\forall I, I' \in \mathfrak{M})\,(I \neq I' \text{ implies } I \not\subseteq I')$.

If $\mathfrak{M} = \{I(\mathcal{A}_j) \mid j = 1, \ldots, r\}$ is a modularization of N, it can be used to define the qualitative dependencies among all attributes of $\mathcal{A}$, which are especially clarified in terms of the induced hypergraphs $(\mathcal{A}, \{\mathcal{A}_1, \ldots, \mathcal{A}_r\})$ and $(N, \mathfrak{M})$, respectively. In order to get a pleasant representation of R_j, $j = 1, \ldots, r$, we apply the concept of the general cartesian product:

Let $I \subseteq N$ be a non-empty index set and $(\Omega_i)_{i \in I}$ its attached set system of domains. Then,

$$\Omega^I = \bigtimes_{i \in I} \Omega_i = \Big\{\varphi \mid \varphi : I \to \bigcup_{i \in I} \Omega_i \quad \text{and} \quad (\forall i \in I)\;(\varphi(i) \in \Omega_i)\Big\}$$

is the product space of $(\Omega_i)_{i \in I}$. Hence, $\Omega \stackrel{\text{Df}}{=} \Omega^N$ denotes the product space of our universe of discourse $\mathfrak{U} = (\Omega_i)_{i \in N}$.

The mentioned quantitative dependencies can now be formalized by a *rule system* $\mathcal{R}(\mathfrak{U}, \mathfrak{M}) = \{R^I \mid I \in \mathfrak{M}\}$, $R^I \subseteq \Omega^I$, such that $R_j = R^{I(\mathcal{A}_j)}$, $j = 1, \ldots, r$.
In a corresponding way the evidential knowledge may be specified with the aid of an *evidential system* $\mathcal{E}(\mathfrak{U}, \mathfrak{N}) = \{E^J \mid J \in \mathfrak{N}\}$, where $\mathfrak{N}$ is a partition of (non-empty subsets of) N, expected to be *compatible with* $\mathfrak{M}$, which means that for all $J \in \mathfrak{N}$ there exists an $I \in \mathfrak{M}$ such that $\emptyset \neq J \subseteq I$. If, for example, only single attribute values, but no relations between attribute values are to be observed, then $\mathfrak{N} = \big\{\{1\}, \ldots, \{n\}\big\}$ is the right choice and always compatible with $\mathfrak{M}$. Note that compatibility ensures that the specific qualitative dependencies in the evidential knowledge coincide with the stated general qualitative dependencies (reflected by the chosen modularization $\mathfrak{M}$).
In order to determine R from $\mathcal{R}(\mathfrak{U}, \mathfrak{M})$ and E from $\mathcal{E}(\mathfrak{U}, \mathfrak{N})$, respectively, and furthermore, to calculate the resulting restrictions for the ${\omega_0}^{(i)}$-values, induced by the assumption that imprecise general and evidential knowledge is *correct* w.r.t. ω_0 (which means that $\omega_0 \in R \cap E$), we need three elementary set-theoretical operations on product spaces, which are *cylindrical extension*, *intersection*, and *projection*.

Let the two non-empty index sets $I, J \subseteq N$ fulfil $I \subseteq J$. Additionally, let $\mathcal{P}(\Omega^I)$ and $\mathcal{P}(\Omega^J)$ be the power sets of Ω^I and Ω^J, respectively.

$$\widehat{\prod}_I^J : \mathcal{P}(\Omega^I) \to \mathcal{P}(\Omega^J),$$

$$\widehat{\prod}_I^J(\Phi) \stackrel{\mathrm{Df}}{=} \Big\{\psi \in \Omega^J \;\Big|\; (\exists \varphi \in \Phi)\,(\forall i \in I)\,\big(\psi(i) = \varphi(i)\big)\Big\}$$

denotes the *cylindrical extension of* Ω^I *onto* Ω^J.

$$\prod_I^J : \mathcal{P}(\Omega^J) \to \mathcal{P}(\Omega^I),$$

$$\prod_I^J(\Psi) \stackrel{\mathrm{Df}}{=} \Big\{\varphi \in \Omega^I \;\Big|\; (\exists \psi \in \Psi)\,(\forall i \in I)\,\big(\psi(i) = \varphi(i)\big)\Big\}$$

is the *projection of* Ω^J *onto* Ω^I.

Using this formalism, we are now in the position to define the exact interpretation of our rule system $\mathcal{R}(\mathfrak{U}, \mathfrak{M})$ and the evidential system $\mathcal{E}(\mathfrak{U}, \mathfrak{N})$ in the way that

$$R \stackrel{\mathrm{Df}}{=} \bigcap_{I \in \mathfrak{M}} \widehat{\prod}_I^N(R^I) \quad \text{is our } \textit{induced general knowledge base}$$

$$\text{and} \quad E \stackrel{\mathrm{Df}}{=} \bigcap_{J \in \mathfrak{N}} \widehat{\prod}_J^N(E^J) \quad \text{our } \textit{induced total evidence w.r.t. } \omega_0 .$$

Note that these definitions consider both, the underlying qualitative dependency structure (in terms of the modularization $\mathfrak{M}$ and the compatible partition $\mathfrak{N}$), and the quantitative knowledge, given by the rule system $\mathcal{R}(\mathfrak{U}, \mathfrak{M})$ and the evidential system $\mathcal{E}(\mathfrak{U}, \mathfrak{N})$.

Regarding the realization of a knowledge propagation mechanism, we want to call $\mathfrak{X} = \big(\mathfrak{U}, \mathfrak{M}, \mathfrak{N}, \mathcal{R}(\mathfrak{U}, \mathfrak{M})\big)$ a *focusing system*, and

$$\sigma\big(\mathfrak{X}, \mathcal{E}(\mathfrak{U}, \mathfrak{N})\big) \stackrel{\mathrm{Df}}{=} R \cap E$$

the *current state of* $\mathfrak{X}$.
The name "focusing system" is due to the fact that we focus our general knowledge $\mathcal{R}(\mathfrak{U}, \mathfrak{M})$ in the light of the available evidential knowledge $\mathcal{E}(\mathfrak{U}, \mathfrak{N})$. The resulting *projections* w.r.t. single attributes are

$$\pi^{(i)} \stackrel{\mathrm{Df}}{=} \prod_{\{i\}}^N \Big(\sigma\big(\mathfrak{X}, \mathcal{E}(\mathfrak{U}, \mathfrak{N})\big)\Big), \; i = 1, \ldots, n.$$

Note that ${\omega_0}^{(i)} \in \pi^{(i)}$ are the most specific restrictions we can get for the attribute values of the object state $\omega_0 = ({\omega_0}^{(1)}, \ldots, {\omega_0}^{(n)})$, assuming that our whole available knowledge on ω_0, represented by $\mathcal{R}(\mathfrak{U}, \mathfrak{M})$ and $\mathcal{E}(\mathfrak{U}, \mathfrak{N})$, is correct w.r.t. ω_0. Furthermore note that the projections $\pi^{(i)}, i = 1, \ldots, n$, are of course less informative than $\sigma\big(\mathfrak{X}, \mathcal{E}(\mathfrak{U}, \mathfrak{N})\big)$, since

$\bigcap_{i=1}^{n} \widehat{\prod}_{\{i\}}^{N} \pi^{(i)} \supseteq R \cap E$, where the equality in general does not hold.

Efficient local propagation techniques that take advantage from the qualitative dependency structure (i.e. the semantics of the modularization $\mathfrak{M}$), are, for example, presented in [Kruse *et al.*, 1994].
General knowledge is often described in terms of *imprecise inference rules* of the form

$$R_j : \textit{ if } \xi^{A_j} \textit{ in } X_j \textit{ then } \xi^{B_j} \textit{ in } Y_j \ ,$$

where ξ denotes a variable taking its values on $\mathcal{P}(\Omega)$, and $\emptyset \neq A_j \subseteq N$, $\emptyset \neq B_j \subseteq N$, $A_j \cap B_j = \emptyset$, $\emptyset \neq X_j \subseteq \Omega^{A_j}$, $\emptyset \neq Y_j \subseteq \Omega^{B_j}$ is supposed to be fulfilled.
For $\emptyset \neq A \subseteq N$, the value ξ^A denotes the Ω^A–projection of ξ.

R_j is interpreted in the way that

$$\begin{array}{lll} \xi^{A_j} \in X_j & \text{implies} & \xi^{B_j} \in Y_j, \quad \text{and} \\ \xi^{A_j} \in \Omega^{A_j} \backslash X_j & \text{implies} & \xi^{B_j} \in \Omega^{B_j}, \end{array}$$

which induces the representing relation

$$R_j = \left(\widehat{\prod}_{A_j}^{I_j}(X_j) \cap \widehat{\prod}_{B_j}^{I_j}(Y_j)\right) \cup \widehat{\prod}_{A_j}^{I_j}(\Omega^{A_j} \backslash X_j)$$

with respect to the common index set $I_j = A_j \cup B_j$.

The occuring system of index sets I_j induces a modularization $\mathfrak{M}$ of N, and therefore, incorporating the relations R_j, a rule system $\mathcal{R}(\mathfrak{U}, \mathfrak{M})$. Hence, for representing our general knowledge, we get a focusing system $(\mathfrak{U}, \mathfrak{M}, \mathfrak{N}, \mathcal{R}(\mathfrak{U}, \mathfrak{M}))$, since only evidence on manifestations is expected to be available.
Realizing imprecise reasoning then means to calculate $\pi^{(i)}$ for all $i \in N$ with respect to a given evidential system $\mathcal{E}(\mathfrak{U}, \mathfrak{N})$.

3 Possibilistic Reasoning

So far we have restricted ourselves to consider a formal reasoning environment that is capable of handling imprecise, but certain knowledge. On the other hand this model can of course be generalized to the treatment of uncertain information. Among the well–known uncertainty calculi for numerical settings such as Bayes theory, Dempster–Shafer theory, possibility theory, and fuzzy set theory, a promising and straight forward extension of our approach refers to possibility theory in the sense of Zadeh [Zadeh, 1978], where possibility distributions are introduced as the epistemic counterparts of fuzzy sets rather than via the definition of possibility measures and necessity measures, respectively. For this reason we prefer to consider *possibility functions* $\pi : \Omega \rightarrow [0,1]$, where $\pi(\omega)$, applied to our reasoning process, quantifies the degree of possibility with which $\omega \in \Omega$ equals the current object state ω_0 of interest. Note that we talk about possibility functions rather than possibility distributions, since in our setting it is more appropriate to reject the normalization assumption and to view π as the one–point coverage of a random set (γ, P), $\gamma : C \rightarrow 2^{\Omega}$ [Nguyen, 1978]. C is assumed to be a finite set of consideration contexts for ω_0, $\gamma(c)$ the specific characterization of ω_0 in context $c \in C$, saying that $\omega_0 \in \gamma(c)$ is true, if c is the true context (which is an event that occurs with probability $P(\{c\})$). With this semantic background π can be defined as $\pi \equiv \pi_\Gamma$, where $\pi_\Gamma : \Omega \rightarrow [0,1]$, $\pi_\Gamma(\omega) \overset{\text{Df}}{=} P(\{c \in C \mid \omega \in \gamma(c)\})$ is fulfilled. Regarding a detailed discussion of possibility functions in a more general view of consideration contexts, including an investigation of reasonable operations on possibilistic data, we refer to [Gebhardt and Kruse, 1993b, Gebhardt and Kruse, 1993c].

Note that $\pi(\omega)$ is the mass of all contexts $c \in C$ that support the possibility of truth of "$\omega = \omega_0$" and therefore serves as a possibility degree. Furthermore note that π_Γ can be interpreted as an information–compressed representation of Γ, restricting to pure possibilistic aspects, avoiding the explicit treatment of underlying consideration contexts.

Based on the random set approach, π_Γ reflects the occurence of uncertainty (probabilistic background in form of the probability measure P) and imprecision (due to the set–valued function γ) for the imperfect characterization of ω_0. In the special case of precision (i.e. $|\gamma(c)| = 1$ for all $c \in C$), π_Γ formally coincides with a probability distribution on Ω, but $\pi_\Gamma(\omega)$ should in general not be mixed up with a probability mass, since — from a semantic point of view — possibility degrees and probability masses are quite different concepts.

With respect to the realization of possibilistic reasoning it has to be pointed out that the whole discussion on focusing systems and evidential systems presented in the previous section can easily be adopted for the possibilistic case in the way that *possibilistic rule systems* $\mathcal{R}(\mathfrak{U}, \mathfrak{M}) = \{\rho^I \mid I \in \mathfrak{M}\}$ and *possibilistic evidential systems* $\mathcal{E}(\mathfrak{U}, \mathfrak{N}) = \{\varepsilon^J \mid J \in \mathfrak{N}\}$ with possibility functions $\rho^I : \Omega^I \rightarrow [0,1]$ and $\varepsilon^J : \Omega^J \rightarrow [0,1]$, respectively, are applied.

In this connection all operations on relations used for the propagation algorithm (which are intersection, cylindrical extension, and projection) have to be replaced by their possibilistic counterparts (e.g. the min-operation for intersection).
Furthermore it turns out that the mentioned context-related view of possibility functions requires to interpret possibilistic inference rules (as a generalization of imprecise inference rules discussed in section 2) with the aid of the corresponding *Gödel relation*. For a clarification of this fact, see, for example, [Gebhardt and Kruse, 1993c].

4 Application and Concluding Remarks

Possibilistic reasoning as introduced in the previous sections, using the concept of a possibilistic focusing system $(\mathfrak{U}, \mathfrak{M}, \mathfrak{N}, \mathcal{R}(\mathfrak{U}, \mathfrak{M}))$, $\mathfrak{U} = (\Omega_i)_{i \in N}$, has been realized in the software tool POSSINFER (Possibilistic Inference) [Kruse *et al.*, 1994], which therefore in some sense is analogous to its probabilistic counterparts HUGIN [Jensen *et al.*, 1990, Andersen *et al.*, 1989] and BAIES [Cowell, 1992].
These tools are implementations of one of the most important algorithms for probabilistic inference [Lauritzen and Spiegelhalter, 1988, Pearl, 1988] and its recent improvements [Lauritzen, 1991, Lauritzen, 1992, Spiegelhalter *et al.*, 1992]. Among other tools that reflect similar approaches for non-Bayesian uncertain reasoning in numerical frameworks, especially PULCINELLA [Saffiotti and Umkehrer, 1991] should be mentioned, which is related to Shenoy's valuation-based systems [Shenoy, 1989, Shenoy and Shafer, 1990], proposing propagation algorithms in terms of valuations that can be particularized to probabilities, beliefs, and boolean expressions, respectively. Selected work that has been done in this field concerns the propagation of upper and lower probabilities in directed acyclic networks [Cano *et al.*, 1991], and corresponding considerations for the belief function setting [Wilson, 1991].
Although propagation in the pure probabilistic environment surely is the most advanced among the numerical approaches (also see, for example, the PATHFINDER project [Heckermann *et al.*, 1992]), it should be recognized that there are also application domains for alternative frameworks, whenever the uncertain knowledge to be modelled is rather imprecise than precise, or, if the specification of prior probabilities and conditional probabilities turns out to be unmotivated or just over-modelled in comparison to the imperfect information that is available in order to

realize an efficient reasoning process. Note that our concept of a possibilistic focusing system and therefore the resulting softwaretool POSSINFER does not explicitly use conditional objects, but only handles the dependencies expressed by possibilistic rule systems $\mathcal{R}(\mathfrak{U},\mathfrak{M})$.

At first glance such an approach may be criticized as being quite restrictive, but in fact there seem to be interesting applications, where this setting is appropriate and sufficient for effective imperfect knowledge modelling.

As an example we considered a causal probabilistic network, which is a directed acyclic graph, consisting of 22 nodes and 22 vertices, for determining the genotype and verifying the parentage of Danish Jersey cattle in the F–blood group system. It is chosen to be a tutorial example for HUGIN and its underlying theoretical background [Rasmussen, 1992]. Presupposing the qualitative dependency structure (in our case the modularization $\mathfrak{M}$ of the possibilistic focusing system that is intended to be developed), we used an available database of 747 cases, each of which specifies the (imprecisely) observed values for 10 of the 22 involved attributes, for finding an appropriate possibilistic rule system $\mathcal{R}(\mathfrak{U},\mathfrak{M})$. More precisely spoken, we defined the random set $\Gamma = (\gamma, P), \gamma : C \rightarrow \mathcal{P}(\Omega), P(\{c\}) \stackrel{\text{Df}}{=} \frac{1}{|C|}$, where C denotes a set of uniformly distributed consideration contexts (interpreted as information sources for the cases in the database), and $\gamma(c)$ the specific (imprecise) datum attached to context $c \in C$. Furthermore let Γ_I denote the random set $\Gamma_I = (\gamma_I, P), \gamma_I : C \rightarrow \mathcal{P}(\Omega^I)$, defined by $\gamma_I(c) = \prod_I^N(\gamma(c)), I \in \mathfrak{M}$. Then, $\mathcal{R}(\mathfrak{U},\mathfrak{M}) = \{\pi_{\Gamma_I} \mid I \in \mathfrak{M}\}$ is the possibilistic rule system induced by the random set Γ and the stated modularization $\mathfrak{M}$.

Related to this example, it turns out that the quality of decisions, delivered by propagation with the aid of the resulting possibilistic focusing system, is similar to that of HUGIN.

Regarding the theoretical background of decision making with possibility functions in the context approach [Gebhardt and Kruse, 1993b], we refer to [Gebhardt and Kruse, 1993c]. Nevertheless a lot of work has still to be done for a well–founded comparative discussion of the two frameworks.

Furthermore we have to investigate algorithms for the appropriate choice of modularizations $\mathfrak{M}$. Corresponding algorithms for the creation of Bayesian belief network structures from data are either based on conditional independence tests (e.g. [Verma and Pearl, 1992]) or take a Bayesian learning method (e.g. [Cooper and Herskovits, 1992, Lauritzen et al., 1993]).

[Singh and Valtorta, 1993] presents an algorithm that integrates these two concepts in order to reduce time complexity. In our future considerations it has to be pointed out, in which way such learning algorithms may be modified and adopted for possibilistic focusing systems.

References

[Andersen *et al.*, 1989] S.K. Andersen, K.G. Olesen, F.V. Jensen, and F. Jensen. HUGIN — A shell for building Bayesian belief universes for expert systems. In *Proc. 11th international joint conference on arificial intelligence*, 1080–1085, 1989.

[Cano *et al.*, 1991] J.E. Cano, S. Moral, and J.F. Verdegay–López. Combination of upper and lower probabilities. In B. D'Ambrosio, P. Smets, and P.P. Bonisonne, editors, *Proc. 7th Conf. on Uncertainty in Artificial Intelligence*, 61–68, San Mateo, 1991. Morgan Kaufmann.

[Cooper and Herskovits, 1992] G. Cooper and E. Herskovits. A Bayesian method for the induction of probabilistic networks from data. *Machine Learning*, 9:309–347, 1992.

[Cowell, 1992] R.G. Cowell. BAIES — A probabilistic expert system shell with qualitative and quantitative learning. In J.M. Bernardo, J.O. Berger, A.P. Dawid, and A.F.M. Smith, editors, *Bayesian statistics 4*. Clarendon Press, Oxford, 1992.

[Dubois and Prade, 1993] D. Dubois and H. Prade. A fuzzy relation–based extension of Reggia's relational model for diagnosis handling uncertain and incomplete information. In *Proc. 9th Conf. on Uncertainty in Artificial Intelligence, Washington*, 106–113, 1993.

[Gebhardt and Kruse, 1993a] J. Gebhardt and R. Kruse. A comparative discussion of combination rules in numerical settings. CEC–ESPRIT III BRA 6156 DRUMS II, Annual Report, 1993.

[Gebhardt and Kruse, 1993b] J. Gebhardt and R. Kruse. The context model — an integrating view of vagueness and uncertainty. *Int. J. of Approximate Reasoning*, 9:283–314, 1993.

[Gebhardt and Kruse, 1993c] J. Gebhardt and R. Kruse. A new approach to semantic aspects of possibilistic reasoning. In M. Clarke, R. Kruse, and S. Moral, editors, *Symbolic and Quantitative Approaches to Reasoning and Uncertainty, Lecture Notes in Computer Science, 747,*, 151–160. Springer, Berlin, 1993.

[Heckermann *et al.*, 1992] D. Heckermann, E. Horvitz, and B. Nathwani. Toward normative expert systems: Part I — The PATHFINDER project. *Methods of Information in Medicine*, 31:90–105, 1992.

[Jensen *et al.*, 1990] F.V. Jensen, B. Chamberlain, and S.K. Andersen. An algebra of Bayesian belief universes for knowledge–based systems. *Networks*, 20:637–659, 1990.

[Kruse *et al.*, 1994] R. Kruse, J. Gebhardt, and F. Klawonn. *Fuzzy Systems, Theory and Applications.* Wiley, Chichester, 1994. (English translation of: Fuzzy Systeme, Series: Leitfäden und Monographien der Informatik, Teubner, Stuttgart, 1993).

[Lauritzen, 1991] S.L. Lauritzen. The EM algorithm for graphical association models with missing data. Technical Report 91-05, Institute for Electronic Systems, Aalborg University, 1991.

[Lauritzen, 1992] S.L. Lauritzen. Propagation of probabilities, means and variances in mixed graphical association models. *J. of the American Statistical Association*, 86, 1992.

[Lauritzen and Spiegelhalter, 1988] S.L. Lauritzen and D.J. Spiegelhalter. Local computations with probabilities on graphical structures and their application to expert systems. *Journal of the Royal Statistical Society, Series B*, 50:157–224, 1988.

[Lauritzen *et al.*, 1993] S.L. Lauritzen, B. Thiesson, and D. Spiegelhalter. Diagnostic systems created by model selection methods — A case study. Preliminary Papers of the 4th Int. Workshop on Artificial Intelligence and Statistics, January 1993.

[Nguyen, 1978] H.T. Nguyen. On random sets and belief functions. *J. of Mathematical Analysis and Applications*, 65:531–542, 1978.

[Pearl, 1988] J. Pearl. *Probabilistic Reasoning in Intelligent Systems: Networks of Plausible Inference.* Morgan Kaufmann, New York, 1988.

[Rasmussen, 1992] L.K. Rasmussen. Blood group determination of Danish Jersey cattle in the F–blood group system. Dina Research Report 8, Dina Foulum, 8830 Tjele, DK, November 1992.

[Reiter, 1987] R. Reiter. Nonmonotonic reasoning. *Annual Review of Computer Science*, 2:147–186, 1987.

[Saffiotti and Umkehrer, 1991] A. Saffiotti and E. Umkehrer. PULCINELLA: A general tool for propagating uncertainty in valuation networks. In B. D'Ambrosio, P. Smets, and P.P. Bonisonne, editors, *Proc. 7th Conf. on Uncertainty in Artificial Intelligence*, 323–331, San Mateo, 1991. Morgan Kaufmann.

[Shenoy and Shafer, 1990] P.P. Shenoy and G.R. Shafer. Axioms for probability and belief–function propagation. In R.D. Shachter, T.S. Levitt, L.N. Kanal, and J.F. Lemmer, editors, *Uncertainty in Artificial Intelligence (4)*, 169–198. North–Holland, Amsterdam, 1990.

[Shenoy, 1989] P.P. Shenoy. A valuation–based language for expert systems. *Int. J. of Approximate Reasoning*, 3:383–411, 1989.

[Singh and Valtorta, 1993] M. Singh and Marco Valtorta. An algorithm for the construction of Bayesian network structures from data. In *Proc. 9th. Conf. on Uncertainty in Artificial Intelligence, Washington*, 259–265, 1993.

[Spiegelhalter *et al.*, 1992] D.J. Spiegelhalter, A.P. Dawid, S.L. Lauritzen, and R.G. Cowell. Bayesian analysis in expert systems. *Statistical Science*, 8:219–247, 1992.

[Verma and Pearl, 1992] T. Verma and J. Pearl. An algorithm for deciding if a set of observed independencies has a causal explanation. In *Proc. 8th Conf. on Uncertainty in Artificial Intelligence*, 323–330, 1992.

[Wilson, 1991] N. Wilson. A montecarlo algorithm for Dempster–Shafer belief. In B. D'Ambrosio, P. Smets, and P.P. Bonisonne, editors, *Proc. 7th Conf. on Uncertainty in Artificial Intelligence*, 414–417, San Mateo, 1991. Morgan Kaufmann.

[Zadeh, 1978] L.A. Zadeh. Fuzzy sets as a basis for a theory of possibility. *Fuzzy Sets and Systems*, 1:3–28, 1978.

4.3. Hybrid Reasoning with FUZZYNEX

Sascha Dierkes, Bernd Reusch, and Karl-Heinz Temme

Abstract

FUZZYNEX is an extension of Nexpert Object to make fuzzy inference available. The knowledge is completely represented within Nexpert Object, no access on external files is done during inference. FUZZYNEX allows the construction of hybrid knowledge-based systems by combination of fuzzy and crisp reasoning.

1 Motivation

The idea for FUZZYNEX origins from a project group [1] at the chair Informatik 1, University of Dortmund, which worked on the optimization of parameters for resistance welding. [2][3].

This optimization was performed by an fuzzy expert system, which supported the user during parameter selection but did not directly control the machine. Based on this problem definition the following constraints for the fuzzy expert system were concluded:

1. representation of fuzzy expert knowledge,
2. data base interfaces to various parameter data bases,
3. representation of crisp knowledge at extreme values,
4. dynamical weights for the knowledge,
5. executable under MS-Windows
6. easy and understandable user manual.

As there was no shell with this desired functionality, we looked for a shell with at least a subset of functions and which was established at the market. Our choice was NEXPERT OBJECT (NO) [10][11][12] under MS-Windows 3.1 together with Toolbook 1.5 by Asymetrix. This

[1]Project groups are mandatory practica for senior students (in groups of 12) for one year at the department Informatik, University of Dortmund. The work of a project group is practical oriented and covers all typical tasks of a software project like brainstorming, conception, implementation, test, documentation etc.

combination satisfied all requests except the fuzzy component for the system. But due to the C-interface of NO a fuzzy component could be integrated.
The project group was successfully finished in March 1993. The idea of FUZZYNEX has been developed further outside university and has changed from a prototype to a product. The status of the system described in this article is of July 1993.

2 Concepts

2.1 General ideas

The important basic idea during the development of the system was the uniform representation of crisp and fuzzy knowledge from the beginning on. The knowledge should be represented completely within NO, so NO had to be enhanced be a fuzzy inference machine, which interprets and evaluates rules in NO notation as fuzzy rules. The base for the fuzzy inference is the common "Fuzzy Logic Control"(FLC) whereby the user can choose between "correlation product inference" or "correlation minimum inference" [6] [8]. For the defuzzification process the center-of-gravity method is used with the modification, that overlapping areas could be either regarded as one single area or each influence the calculation.

2.2 Hypothesis-based inference

In addition to the FLC-inference another concept, which is a real extension, is implemented. This concept focusses on inferences which are based on hypothesises (in short: hypos). The inference machine receives hypos and tries to validate them by rules, in case of crisp logic these hypos can be either true or false, but in FUZZYNEX the membership value ($\mu_{\mathcal{H}_w}$(Hypothese)) of a hypo in respect to the set of valid hypos ($\mathcal{H}_w$) is being calculated. The crisp logic is here a special case, if the hypo is true, its membership value in respect to the set of valid hypos is 1, otherwise 0 (false hypo). The membership value of a hypo to $\mathcal{H}_w$is determined by an OR-operation of the membership values of the rules in respect to the set of valid rules ($\mathcal{R}_w$, or $\mu_{\mathcal{R}_w}$(rule)). The membership value of a rule to $\mathcal{R}_w$is then determined by an AND-Operation of the membership values of the rule's premises in respect to the set of valid premises ($\mathcal{P}_w$, or $\mu_{\mathcal{P}_w}$(premise)) multiplied by the factor for the certainty of the rule (*Certainty factor*). The algorithm looks like:

$$\mu_{\mathcal{H}_w}(\text{hypothesis H}) = \mu_{\mathcal{R}_w}(\text{rule}_1 \text{ (H)}) \vee_f \cdots \vee_f \mu_{\mathcal{R}_w}(\text{rule}_n \text{ (H)}) \quad (1)$$

$$\mu_{\mathcal{R}_w}(\text{rule R}) = (\mu_{\mathcal{P}_w}(\text{premise}_1(\text{R})) \wedge_f \cdots \mu_{\mathcal{P}_w}(\text{premise}_n(\text{R}))) * certaintyfactor(R) \quad (2)$$

The membership factor of a premise to the set of valid premises is calculated depending on the premise. In the trivial case it is a membership value of a fuzzy set:
Is age "young".
In this example the membership value of the variable *age* in respect to the fuzzy set *young* is calculated, which is then the value for $\mu_{\mathcal{P}_w}$(Is age "young"). The second non-trivial case for premises is the backward chaining of hypos, for example
Yes age-is-young.
In this case is $\mu_{\mathcal{P}_w}$(Yes age-is-young) $= \mu_{\mathcal{H}_w}(\text{age} - \text{is} - \text{young})$.
This value is calculated by rules as mentioned above.

In parallel rules may contain crisp premises, these premises are used to decide whether a rule should be processed further or not.
The ability to combine crisp and fuzzy premises and to connect rules by hypos is a good opportunity to construct structured fuzzy rule networks. So at start it could be determined by a crisp premise, whether a rule should be fired or not, then a particular rule set could be processed via the hypo connection.
A second alternative is to control rules by the dynamical *certainty factor* of each rule. If it is set to 0, the conclusion of the rule is not evaluated.

2.3 Connection of crisp and fuzzy inference

Within the system the fuzzy inference is totally seperated from the crisp inference of NO. The fuzzy inference is called by an execute-handler, to which those hypos are passed, which should be fuzzy evaluated. This offers the opportunity to select dedicated rule classes for inference. Hypos can be bound dynamically on a class which is then transferred to the execute-handeler to be evaluated.

3 Functionality

3.1 Standard FLC inference

At first FUZZYNEX offers the common functionality which is used in many fuzzy controllers. But the system is not designed to program fuzzy controllers but for applications in the field of information and advising systems. It is bound for knowledge engineers of expert systems, who intend to solve their problems with the aid of as well crisp as fuzzy knowledge. NO has been established as the leading tool on market for processing crisp knowledge, now FUZZYNEX offers the opportunity to handle fuzzy knowledge under NO, too.

In advising systems which should reflect expert knowledge it is frequently necessary to model complex decision processes [1]. To support this modelling FUZZYNEX offers hypo-based inference to structure the knowledge. This can be useful even for rule networks with only one stage. For example during the project group already mentioned above an one-stage rule base with about 170 fuzzy rules was implemented. Due to the flat structure correction and test of the rules was difficult. After a redesign and reimplementation as a multi-stage knowledge base these points were much easier to perform.

3.2 Fuzzy editor

To work correctly certain requirements were made for the inference. These interface definitions are uncomfortable for the user, so a graphical surface under MS-Windows 3.1 with Toolbook 5.1 was developed for the creation of fuzzy data bases. This surface is designed not to allow the user to give wrong input, another feature is the graphical display and editing of fuzzy sets.

3.3 Combination of crisp and fuzzy reasoning

FUZZYNEX offers the opportunity to connect crisp and fuzzy knowledge. The first or "top" rule always has to be a crisp one in NO, from which the fuzzy inference process can be started. To design a hybrid knowledge base and to use both types of knowledge, several alternatives can be used. The crisp inference can influence the fuzzy one as follows:

from	to	methods
crisp	crisp	standard inference of NO
crisp	fuzzy	• determination by crisp rules whether a fuzzy inference should be started or not, • detemination by hypos, which rule branches should be started, • dynamical assignment of *certainty factors.*
fuzzy	crisp	not possible
fuzzy	fuzzy	connection of rules by hypos (in combination with meta-premises: selection of fuzzy rules during inference)

Table 1: Overview combinations of inference strategies

- depending on rules the fuzzy inference can be started,
- class structures can be used to determine which hypos should be evaluated [2],
- out of a slot the confidence factor of a rule can be set.

Besides the control of the fuzzy inference by the crisp inference the opportunity of hybrid rules exists, which combine crisp and fuzzy premises. The crisp premises then play the role of *meta-premises.* These meta-premises decide whether a rule is passed to the fuzzy inference or not. When a crisp premise occurs within a hybrid rule, the evaluation is interrupted. If the premise is false, the evaluation is aborted and the rule does not contribute to the calculation of $\mu_{\mathcal{H}_w}()$ of the appropriate hypo. By using these meta-premises it is very easy to activate or deactivate particular rule branches according to certain crisp facts. For this it is necessary that those facts are already known, there is no dynamical determination of the value of a fact. This would be a connection from the fuzzy to the crisp inference and is not yet designed in the system.
1

[2]caused by so called "dynamic links"

4 Implementation

FUZZYNEX is an extension of NO which is implemented in ANSI-C using the "Application Programming Interface" (API) of NO. It consists of one program for the inference and one program for the defuzzification. Both programs are directly incorporated in NO and do not access any external information. All rules, objects and fuzzy sets are completely represented within NO and can be edited by the build-in surface of NO. They have to obey certain syntactic and semantic restrictions, e.g. for the inference a particular class-object-structure must exist to represent the fuzzy variables and fuzzy sets.

4.1 Fuzzy frames in NO

The representation of variables for the fuzzy inference is implemented as objects in NO, which belong to the class *Fuzzy_Objects* and have the following slots:

1. *x_Minimum* resp. *x_Maxiumum*, these slots hold the range for which the variable is defined [3],
2. *Unit*, unit for the x-range,
3. *Comment*, slot for explanation text,
4. *FuzzyAttributes*, list of attributes which are defined for this variable,
5. *FuzzyObject_Source*, reference for the crisp value to work with during inference,
6. *Area*, size of area which was calculated during defuzzification,
7. *center of gravity*, x-coordinate of the cog,

Fuzzy sets are represented as objects in NO, too. The sets are defined by a finite number of values, between these samples the functions are linear. For this the objects have two slots to store the coordinates of the samples as lists. A third slot is reserved for comments.
Another class of objects is provided for the fuzzy hypos. In one (boolean) slot it is marked whether the hypo has been already processed, a second slot takes the membership value of the hypo in respect to the set of all valid hypos after the hypo has been processed.

[3] range of universe

4.2 Incorporation into NO

NO offers an interface via API through which all functions of NO can be activated and all information of a data base can be read or written. The programs for inference and defuzzification access via this interface the data base for the fuzzy rules. Within NO the programs are called by an 'execute' statement which is placed in a slot or metaslot. A limited set of operaters in NO is offered for use by rules, concerning the inference these are:

IS calculates the membership value, syntax is:
IS FuzzyObject attribute

an example:
IS FuzzyObject_person "young"

ISNOT is equivalent to Fuzzy-Not(*IS*)

YES calculates the membership value of a hypo to the set of all valid hypos and starts a backward-chained inference, syntax is:
YES FuzzyHypo

for example:
YES FuzzyHypo_AgeIsYoung

NO is equivalent to Fuzzy-Not(*YES*).

4.3 Inference

The Inference is called by:
Execute "eval" Parameter
The first parameter is a list of hypos which should be evaluated. Further parameters can be adjusted:

- Operator for the combination of premises, e.g. Minimun, Gamma-Operator or Yager-family,
- Operator for the inference (Minimum or Product),
- Operator for the combination of rules, e.g. Maxiumum, Gamma-Operator or drastic sum,
- Operator for NOT,

- control comments, put into Transcript.

During the inference new fuzzy sets are generated as dynamical objects due to the evaluation of the rules, and the $\mu_{\mathcal{H}_w}()$ of the hypos are determined.

4.4 Defuzzification

In the defuzzification process the dynamically generated fuzzy sets are processed by the center-of-gravity method. A modification is available, all sets can first be joint by maximum, or each area counts for its own. The defuzzification routine is called by:
Execute "defuzzify" Parameter
with a list of variables to be defuzzified and the mode of the defuzzification method.

4.5 Hardware platform

FUZZYNEX is available for MS-Windows 3.1 and for VAX/VMS. For the MS-Windows 3.1 version a Fuzzy Editor is offered too, which supports the creation of a fuzzy knowledge base.

4.6 Perspectives

FUZZYNEX has been developed as an aid to create hybrid knowledge bases. Further developments will concentrate on structuring those bases. Chaining rules is one more field of investigation [4] as well as non standard inference methods [4][5][9].

References

[1] Bernd Reusch (Hrsg.). *Fuzzy Logic: Theorie und Praxis.* Springer Verlag Berlin Heidelberg New York, 1993.

[2] K.-H. Temme. Einsatz von Fuzzy Methoden beim Widerstandspunktschweißen. Forschungsbericht Nr. 449, FB Informatik, Universität Dortmund, 1992. 12. Workshop Interdisziplinäre Methoden in der Informatik.

[4] currently only backward chaining by hypos

[3] PG 210 (Eifer). Ein Fuzzy-Expertensystem zur Maschinenparameterbestimmung. Forschungsberichte Nr. 426 und 486, FB Informatik, Universität Dortmund, 1992/93.

[4] H. Kern, H. Polrolnizcak, K.-H. Temme. Abschlußbericht des BIP-Projektes. Technical report, FB Maschinenbau/Informatik, Universität Dortmund, Schweißtechnische Lehr- und Versuchsanstalt Duisburg, erscheint demnächst.

[5] M. Fathi, K.-H. Temme. Parameteroptimierung am Beispiel des Widerstandpunktschweißens durch ein Fuzzy Expertensystem. In *Konferenzband zur VDI-Tagung "Kosten senken durch EDV-Anwendungen in der Werkstofftechnik"*, Düsseldorf, 1993.

[6] B. Kosko. *Neural Networks and Fuzzy Systems.* Prentice Hall, Englewood Cliffs, N.J., 1992.

[7] A. Kandel, editor. *Fuzzy Expert Systems.* CRC Press, Boca Raton, 1991.

[8] J. Kahlert, H. Frank. *Fuzzy-Logik und Fuzzy-Control.* Vieweg, Braunschweig/Wiesbaden, 1993.

[9] D. Dubois, H. Prade. Fuzzy sets in approximate reasoning, part 1: Inference with possibility distributions. *Fuzzy Sets and Systems*, 40, 1991.

[10] Neuron Data. *Nexpert Object: Functional Description Manual*, 1991.

[11] Neuron Data. *Nexpert Object: User's Guide*, 1991.

[12] Neuron Data. *Nexpert Object: API Reference Manual*, 1991.

4.4. The Semantics of Imprecision in Terminological Logics

Jochen Heinsohn

Abstract

This paper presents the language $\mathcal{ALCP}$ which is a probabilistic extension of terminological logics and aims at closing the gap between *terminological knowledge representation* and *uncertainty handling*. We present the formal semantics underlying the language $\mathcal{ALCP}$ and introduce the probabilistic formalism that is based on classes of probabilities and is realized by means of probabilistic constraints. Besides infering implicitly existent probabilistic relationships, the constraints guarantee terminological and probabilistic consistency. Altogether, the new language $\mathcal{ALCP}$ applies to domains where both term descriptions and uncertainty have to be handled.

1 Introduction

Research in knowledge representation led to the development of terminological logics [Nebel, 1990] which originated mainly in Brachman's KL-ONE [Brachman and Schmolze, 1985] and are called *description logics* [Patil *et al.*, 1992] since 1991. In such languages the terminological formalism (*TBox*) is used to represent a hierarchy of terms (*concepts*) that are partially ordered by a subsumption relation: concept B is *subsumed by* concept A, if, and only if, the set of B's real world objects is necessarily a subset of A's world objects. In this sense, the semantics of such languages can be based on set theory. Two-place relations (*roles*) are used to describe concepts. In the case of *defined* concepts, restrictions on roles represent both necessary and sufficient conditions. For *primitive* concepts, only necessary conditions are specified. The algorithm called *classifier* inserts new generic concepts at the most specific place in the terminological hierarchy according to the subsumption relation. Work on terminological languages led further to *hybrid* representation systems.

This work was supported by the German Ministry for Research and Technology (BMFT) under contract ITW 8901 8 as part of the WIP project. I would like to thank Bernhard Nebel for valuable comments on earlier versions of this paper.

Systems like BACK, CLASSIC, LOOM, KANDOR, KL-TWO, KRIS, KRYPTON, LILOG, MESON, SB-ONE, and YAK (for overview and analyses see [Sigart Bulletin, 1991, Heinsohn *et al.*, 1994]) make use of a separation of terminological and assertional knowledge.

Since, on one hand, the idea of terminological representation is essentially based on the possibility of *defining* concepts (or at least specifying necessary conditions), the classifier can be employed to draw correct inferences. On the other hand, characterizing domain concepts only *categorically* can lead to problems, especially in domains where certain important properties cannot be used as part of a concept definition. This may happen especially in real world applications where, besides their description, terms can only be characterized as having additional *typical* properties or properties that are, e.g., *usually* true. In the real world such properties often are only *tendencies*. Until now, imprecision of this kind has not been considered in the framework of terminological logics.

While, as argued above, classical *terminological knowledge representation* excludes the possibility to handle uncertain concept descriptions, purely numerical *approaches for handling uncertainty* [Kruse *et al.*, 1991] in general are unable to consider terminological knowledge. The basic idea underlying the formalism presented in this paper is to generalize terminological logics by using probabilistic semantics and in this way to close the gap between the two mentioned areas of research.

This paper presents the semantics of the language $\mathcal{ALCP}$ [Heinsohn, 1993] and pursues our earlier investigations [Heinsohn, 1991]. First, we briefly introduce $\mathcal{ALC}$ [Schmidt-Schauß and Smolka, 1991], a propositionally complete terminological language containing the logical connectives conjunction, disjunction and negation, as well as role quantification. In Section 3 we extend $\mathcal{ALC}$ by defining syntax and semantics of *probabilistic conditioning* (p-conditioning), a construct aimed at considering uncertain knowledge sources and based on a statistical interpretation. In Section 4 we introduce the formal model underlying both the terminological and the probabilistic formalism. We further characterize the classes of probabilities induced by terminology and p-conditionings. As demonstrated in Section 5, a set of consistency requirements have to be met on the basis of terminological and probabilistic knowledge. Moreover, the developed interval-valued probabilistic constraints allow the inference of implicitly existent probabilistic relationships and their quantitative computation. The conclusions are given in Section 6. While this paper mainly focuses on terminological and probabilistic aspects, the consideration of *individuals* means the ability to draw inferences about "probabilistic memberships". The complete model based on probabilities over both domains and worlds is described in [Heinsohn, 1993].

2 The Terminological Formalism

The basic elements of the terminological language $\mathcal{ALC}$ [Schmidt-Schauß and Smolka, 1991] are concepts and roles (denoting subsets of the domain of interest and binary relations over this domain, respectively). Assume that $\top$ ("top", denoting the entire domain) and $\bot$ ("bottom", denoting the empty set) are concept symbols, that A denotes a concept symbol, and R denotes a role. Then the concepts (denoted by letters C and D) of the language $\mathcal{ALC}$ are built according to the abstract syntax rule

$$C, D \rightarrow A \mid \top \mid \bot \mid C \sqcap D \mid C \sqcup D \mid \neg C \mid \forall R{:}C \mid \exists R{:}C$$

where $\sqcap$, $\sqcup$, and $\neg$ denote concept conjunction, disjunction, and negation, and $\forall$ and $\exists$ denote value and existential restriction, respectively.

With an introduction to formal semantics of $\mathcal{ALC}$ in mind, we give a translation into set theoretical expressions with $\mathcal{D}$ being the domain of discourse. For that purpose, we define a mapping $\mathcal{E}$ that maps every concept description to a subset of $\mathcal{D}$ and every role to a subset of $\mathcal{D} \times \mathcal{D}$ in the following way:

$$\begin{aligned}
\mathcal{E}[\![\top]\!] &= \mathcal{D}, \quad \mathcal{E}[\![\bot]\!] = \emptyset, \quad \mathcal{E}[\![(\neg C)]\!] = \mathcal{D} \setminus \mathcal{E}[\![C]\!], \\
\mathcal{E}[\![(C \sqcap D)]\!] &= \mathcal{E}[\![C]\!] \cap \mathcal{E}[\![D]\!], \quad \mathcal{E}[\![(C \sqcup D)]\!] = \mathcal{E}[\![C]\!] \cup \mathcal{E}[\![D]\!], \\
\mathcal{E}[\![(\forall R : C)]\!] &= \{x \in \mathcal{D} \mid \text{for all } y \in \mathcal{D} : ((x,y) \in \mathcal{E}[\![R]\!] \Rightarrow y \in \mathcal{E}[\![C]\!])\}, \\
\mathcal{E}[\![(\exists R : C)]\!] &= \{x \in \mathcal{D} \mid \text{exists } y \in \mathcal{D} : ((x,y) \in \mathcal{E}[\![R]\!] \wedge y \in \mathcal{E}[\![C]\!])\}.
\end{aligned}$$

Concept descriptions are used to state necessary, or necessary and sufficient conditions by means of specializations "$\sqsubseteq$" or definitions "$\doteq$", respectively. Assuming symbol A and concept description C, then "$A \sqsubseteq C$" means the inequality $\mathcal{E}[\![A]\!] \subseteq \mathcal{E}[\![C]\!]$, and "$A \doteq C$" means the equation $\mathcal{E}[\![A]\!] = \mathcal{E}[\![C]\!]$. A set of well formed concept definitions and specializations forms a *terminology*, if every concept symbol appears at most once on the left hand side and there are no terminological cycles.

Definition 1 *Let $\mathcal{T}$ be a terminology. The set*

$$\text{mod}(\mathcal{T}) \stackrel{\text{def}}{=} \{\mathcal{E} \mid \mathcal{E} \textit{ extension function of } \mathcal{T}\} \tag{1}$$

is called set of models of $\mathcal{T}$.

A concept C_1 is said to be ***subsumed by*** a concept C_2 in a terminology $\mathcal{T}$, written $C_1 \preceq_{\mathcal{T}} C_2$, iff the inequality $\mathcal{E}[\![C_1]\!] \subseteq \mathcal{E}[\![C_2]\!]$ holds for all extension functions satisfying the equations introduced in $\mathcal{T}$ (i.e., for all $\mathcal{E} \in \mathrm{mod}(\mathcal{T})$).

Terminological languages as $\mathcal{ALC}$ can be usefully applied to *categorical* world knowledge. For instance, we may introduce

Example 1

$$
\begin{array}{rcl}
\textit{animal} & \sqsubseteq & \top \\
\textit{flying} & \sqsubseteq & \top \\
\textit{antarctic\_animal} & \sqsubseteq & \textit{animal} \\
\textit{bird} & \doteq & \textit{animal} \sqcap (\forall \textit{moves\_by} : \textit{flying}) \\
\textit{antarctic\_bird} & \doteq & \textit{antarctic\_animal} \sqcap \textit{bird} \\
\textit{penguin} & \sqsubseteq & \textit{antarctic\_bird}
\end{array}
$$

However, imprecise information cannot be expressed and used in classical terminological logics. The importance of having appropriate language constructs becomes obvious, if we examine the above birds' taxonomy in more detail: Because of terminological subsumption, the flying property of birds is inherited also to the penguin concept. However, it is well known that concerning this aspect penguins represent a real exception, so that the (categorical) definition of birds seems also to be inadequate: At best "*most* birds move by flying" or are flying objects that are defined to move by flying. It seems to be more suitable to generally consider the "degree of intersection" between the respective concept's extensions and to characterize it using an appropriate technique. The idea behind this generalization is to use probabilistic semantics.

3 Probabilistic Conditioning

In the following we consider only one representative for equivalent concept expressions (such as A, $A \sqcap \top$, $A \sqcap A$). The algebra based on representatives of equivalence classes and on the logical connectives $\sqcap$, $\sqcup$, and $\neg$ is known as *Lindenbaum algebra* of the set $\mathcal{S}$ of concept symbols. We use the symbol $\mathcal{C}$ for the set of concept descriptions. Domain $\mathcal{D}$ is assumed to be finite. As a language construct that takes into account *overlapping* concept extensions, we introduce the notion of *p-conditioning*: the language construct $C_1 \overset{[p_l,p_u]}{\longrightarrow} C_2$ is called p-conditioning, iff $[p_l, p_u]$ is a subrange of real numbers with $0 \leq p_l \leq p_u \leq 1$ and $C_1, C_2 \in \mathcal{C}$. The semantic is defined as follows:

Definition 2 *An extension function $\mathcal{E}$ over $\mathcal{C}$ satisfies a p-conditioning $C_1 \stackrel{[p_l,p_u]}{\longrightarrow} C_2$, written $\models_{\mathcal{E}} C_1 \stackrel{[p_l,p_u]}{\longrightarrow} C_2$, iff $|\mathcal{E}[\![C_1 \sqcap C_2]\!]|/|\mathcal{E}[\![C_1]\!]| \in [p_l, p_u]$ holds for concepts $C_1, C_2 \in \mathcal{C}$, $\mathcal{E}[\![C_1]\!] \neq \emptyset$.*

From the above it is obvious that we use the *relative cardinality* for interpreting the notion of p-conditioning. For illustrating the meaning of Definition 2, assume that an observer examines the flying ability of birds in more detail. When finishing his study he may have learned that, different from the model of Example 1, relation *moves_by:flying* holds only for a certain percentage of the birds. The notion of p-conditioning now allows a representation of universal knowledge of statistical kind in a way that maintains the semantics of the roles: the new concept *flying_object* is created with role *moves_by* restricted to range *flying*. The uncertainty is represented by a p-conditioning stating that "at least 95% of *birds* are *flying_objects* that, by definition, all move by flying". The now more detailed view to the example world leads to the following revision of Example 1:

Example 2 *animal* $\sqsubseteq \top$, *flying* $\sqsubseteq \top$, *flying_object* $\doteq \forall$*moves_by* : *flying*, *antarctic_animal* $\sqsubseteq$ *animal*, *bird* $\sqsubseteq$ *animal*, *antarctic_bird* $\doteq$ *antarctic_animal* $\sqcap$ *bird*, *penguin* $\sqsubseteq$ *antarctic_bird*, *bird* $\stackrel{[0.95,1]}{\longrightarrow}$ *flying_object*, *bird* $\stackrel{[0.20,0.20]}{\longrightarrow}$ *antarctic_bird*, *penguin* $\stackrel{[0,0]}{\longrightarrow}$ *flying_object*.

This demonstrates that set theory is sufficient for a consistent semantic basis on which both terminological and probabilistic language constructs can be interpreted. On this basis, the p-conditioning serves also as a generalization of both "inclusion" and "disjointness" (now appearing as $A \stackrel{1}{\rightarrow} B$ and $A \stackrel{0}{\rightarrow} B$, respectively). Example 2 shows not only an adequate representation of the fact that "most (i.e., $\geq$ 95%) birds are flying objects" but also that "20% of the birds are antarctic birds" and "no penguin is a flying object". This directly leads to the question in which way inferences can be drawn on the basis of terminological and probabilistic knowledge to infer implicitly existent relationships. In fact, Example 2 implicitly covers the knowledge, that "at least 75% of antarctic birds are flying objects" and that "at most 5% of birds are penguins", for instance. For this, we first introduce the formal model based on classes of probabilities and then derive the associated probabilistic constraints.

4 The Formal Model

In concrete application domains, knowledge about uncertain concept relations generally exists only for some pairs of concepts of a terminology—neither directly representable statistical knowledge nor textbook knowledge is complete in this sense. Consequently, the question arises in which way, starting with a set of models restricted wrt. terminology and p-conditionings, one can infer (uncertain) relationships between those pairs of concepts for which p-conditionings are not explicitly introduced. Below we give an answer to this question by defining the sets of *entailed* and *minimal* p-conditionings. The first part of the definition considers the fact that the set of models of a terminology (see equation (1)) is generally refined if p-conditionings are introduced.

Definition 3 *Let $\mathcal{T}$ be a terminology and $\mathcal{I}$ be a set of p-conditionings.*

$$\mathrm{mod}_{\mathcal{T}}(\mathcal{I}) \stackrel{\mathrm{def}}{=} \{\mathcal{E} \mid \models_{\mathcal{E}} \mathcal{I}\} \cap \mathrm{mod}(\mathcal{T}) \tag{2}$$

is called the set of models of $\mathcal{I}$ wrt. $\mathcal{T}$.

$$\mathrm{Th}_{\mathcal{T}}(\mathcal{I}) \stackrel{\mathrm{def}}{=} \{I \mid \textit{for all } \mathcal{E} \in \mathrm{mod}_{\mathcal{T}}(\mathcal{I}) : \models_{\mathcal{E}} I\} \tag{3}$$

is called the set of entailed p-conditionings wrt. $\mathcal{I}$ and $\mathcal{T}$.

$$\mathrm{min}_{\mathcal{T}}(\mathcal{I}) \stackrel{\mathrm{def}}{=} \bigcup_{C,D \in \mathcal{C}} \{C \stackrel{R_{\min}}{\rightarrow} D \mid R_{\min} = \bigcap_{R \,:\, C \stackrel{R}{\rightarrow} D \in \mathrm{Th}_{\mathcal{T}}(\mathcal{I})} R\} \tag{4}$$

is called the set of minimal p-conditionings wrt. $\mathcal{I}$ and $\mathcal{T}$.

These definitions—especially the set defined in (4)—describe a formal model that characterizes the computation of p-conditionings introduced not explicitly and the further refinement of p-conditionings that are known. Note that both sets (3) and (4) contain p-conditionings for *all* pairs of concepts. While (3) generally is of infinite size, (4) contains exactly one p-conditioning for one pair of concepts. A set $\mathcal{I}$ of p-conditionings is called *consistent wrt.* $\mathcal{T}$, iff $\mathrm{mod}_{\mathcal{T}}(\mathcal{I}) \neq \emptyset$ holds.

Definition 4 *A concept C_1 is said to be subsumed by a concept C_2 wrt. $\mathcal{T}$ and $\mathcal{I}$, written $C_1 \preceq_{\mathcal{T},\mathcal{I}} C_2$, iff the inequality $\mathcal{E}[\![C_1]\!] \subseteq \mathcal{E}[\![C_2]\!]$ holds for all extension functions $\mathcal{E} \in \mathrm{mod}_{\mathcal{T}}(\mathcal{I})$.*

It can be shown that all minimal sets $R_{\min}$ of real numbers defined in (4) form ranges as it is the case for explicitly introduced p-conditionings. This is due to the convexity property of those probability classes that are induced by terminological axioms and p-conditionings over the set of atomic concept expressions. In the following we focus on this aspect.

In addition to the symbol $\mathcal{C}$ for the set of concept descriptions we use $\mathcal{C}^{\mathrm{A}}$ for the set of *atomic concept expressions* (i.e., the atoms of the Lindenbaum algebra). Atomic concept expressions are of the form $B_1 \sqcap B_2 \sqcap \ldots \sqcap B_m$, where B_i is either a concept symbol A or the negation $\neg A$ of a symbol. The relation $\mathcal{C}^{\mathrm{A}} \subseteq \mathcal{C}$ holds. A first simple observation is that for every extension function $\mathcal{E} \in \mathrm{mod}_{\mathcal{T}}(\mathcal{I})$ the set of extensions of the elements in $\mathcal{C}^{\mathrm{A}}$ forms a partition of $\mathcal{D}$. A direct consequence of this observation is that every extension function $\mathcal{E}$ uniquely determines a probability over $\mathcal{C}^{\mathrm{A}}$.

Proposition 1 *Let $\mathcal{T}$ be a terminology and $\mathcal{I}$ be a consistent set of p-conditionings. Further, let $\mathcal{E} \in \mathrm{mod}_{\mathcal{T}}(\mathcal{I})$ be an extension function for which*

$$\models_{\mathcal{E}} \top \overset{p_i}{\to} C_i^-, \quad p_i \overset{\mathrm{def}}{=} \frac{|\mathcal{E}[\![C_i^-]\!]|}{|\mathcal{D}|} \qquad \textit{for all } C_i^- \in \mathcal{C}^{\mathrm{A}} \tag{5}$$

holds. Then the real-valued set function $P_{\mathcal{E}}$ defined by

$$P_{\mathcal{E}} : 2^{\mathcal{C}^{\mathrm{A}}} \to [0,1],\ P_{\mathcal{E}}(\{C_i^-\}) \overset{\mathrm{def}}{=} p_i,\ C_i^- \in \mathcal{C}^{\mathrm{A}} \tag{6}$$

is a probability function over $\mathcal{C}^{\mathrm{A}}$.

Note that every concept can be represented as a disjunction of atomic concept expressions, i.e., for every concept expression $C \in \mathcal{C}$ there exists a subset $D \subseteq \mathcal{C}^{\mathrm{A}}$ of atoms such that $C = \bigsqcup D$. In this way $P_{\mathcal{E}}$ can be extended to concept expressions.

Proposition 1 shows that, assuming complete knowledge of domain $\mathcal{D}$ and of the involved cardinalities, a probability function $P_{\mathcal{E}}$ over $\mathcal{C}^{\mathrm{A}}$ is induced by the extension function $\mathcal{E}$. However, it is generally more realistic to assume less complete knowledge and cardinalities that are rather relative. Consequently, the set $\mathrm{mod}_{\mathcal{T}}(\mathcal{I})$ generally contains more than one element, so that a *class of probabilities* is induced by $\mathcal{T}$ and $\mathcal{I}$. The most general set of all probabilities over $\mathcal{C}^{\mathrm{A}}$ is defined by

$$\mathcal{M} \overset{\mathrm{def}}{=} \{(p_1, \ldots, p_n) \in \Re^n \mid p_1 + \ldots + p_n = 1,\ p_i \geq 0 \text{ for all } 1 \leq i \leq n\},$$

with $n = 2^m$. Without any knowledge about p-conditionings and terminology the set $\mathcal{M}$ characterizes the status of *complete ignorance*. On the other hand, for a particular extension function $\mathcal{E}$ the set $\mathcal{M}$ consists of exactly one point in the n-dimensional space $[0,1]^n$. In case of given $\mathcal{T}$ and $\mathcal{I}$, by (2) a *set* $\text{mod}_\mathcal{T}(\mathcal{I})$ of extension functions and also a set

$$\begin{aligned}\mathcal{M}_{\mathcal{T},\mathcal{I}} \stackrel{\text{def}}{=} & \{(p_1,\ldots,p_n) \in \mathcal{M} \mid \\ & \text{exists } \mathcal{E} \in \text{mod}_\mathcal{T}(\mathcal{I}) : p_j = P_\mathcal{E}(\{C_j^-\}),\ j = 1,\ldots,n\}\end{aligned}$$

of probabilities are defined. $\mathcal{M}_{\mathcal{T},\mathcal{I}}$ corresponds to the set of probabilities in $\mathcal{M}$ that are compatible with $\mathcal{T}$ and all p-conditionings in $\mathcal{I}$. It can be shown that for every consistent $\mathcal{T}$ and $\mathcal{I}$ $\mathcal{M}_{\mathcal{T},\mathcal{I}}$ is a convex set.

5 Probabilistic Constraints

In the following, we focus on *probabilistic constraints* which correspond to the formal model introduced above, which are *locally* defined and therefore *context-related*, and which *derive* and *refine* p-conditionings and check in this way the consistency of the knowledge base. While in this paper we restrict ourselves to simple examples, the complete set of constraints is given in [Heinsohn, 1993]. We assume consistent sets $\mathcal{T}$ and $\mathcal{I}$ for the rest of this paper.

Proposition 2 *For all concepts $C, D \in \mathcal{C} \setminus \{\bot\}$:*

$$D \preceq_{\mathcal{T},\mathcal{I}} C \;\Leftrightarrow\; \left(D \xrightarrow{1} C\right) \in \min_\mathcal{T}(\mathcal{I}) \quad (7)$$

$$\{C \xrightarrow{[p_l,p_u]} D, D \xrightarrow{[q_l,q_u]} C\} \subseteq \min_\mathcal{T}(\mathcal{I}) \;\Rightarrow\; (p_l > 0 \Leftrightarrow q_l > 0) \quad (8)$$

These constraints characterize the relations between subsumption and p-conditioning ((7)) and focus on the role of disjointness ((8)). In the following we concentrate on *triangular cases* that take into account three concept expressions and allow the inference of minimal p-conditionings. Note that the following proposition examines *primitive* concepts. If a subsumption relation between concepts is known, the respective p-conditioning has to have the range [1, 1] (compare (7)).

Proposition 3 *Assume concepts A, B, C, and p-conditionings*

$$\mathcal{I} = \{A \xrightarrow{[p_l,p_u]} C, A \xrightarrow{[q_l,q_u]} B, B \xrightarrow{[q_l',q_u']} A, C \xrightarrow{[p_l',p_u']} A, p_l' = 0, q_l \neq 0\}.$$

Then $B \overset{R_{\min}}{\rightarrow} C \in \min_{\mathcal{T}}(\mathcal{I})$ *has the minimal range*

$$R_{\min} = \left[\frac{q_l'}{q_l} \cdot \max(0, q_l + p_l - 1), \min\left(1,\ 1 - q_l' + p_u \cdot \frac{q_l'}{q_l}\right)\right]. \quad (9)$$

Note that because of assumption $p_l' = 0, q_l \neq 0$ we restrict ourselves to a special case that however applies to the first situation discussed at the end of Section 3.

While above proposition covers cases, in which only *primitive* concepts are involved, in the case of logically interrelated concepts probabilistic constraints have to be further strengthened to guarantee the minimality of ranges. Assuming $A, B, C \in \mathcal{C} \setminus \{\bot\}$ an example is

$$\left(A \overset{[p_l,p_u]}{\rightarrow} (A \sqcap B)\right) \in \min_{\mathcal{T}}(\mathcal{I}) \Leftrightarrow \left(A \overset{[1-p_u,1-p_l]}{\rightarrow} (A \sqcap \neg B)\right) \in \min_{\mathcal{T}}(\mathcal{I}).$$

The main advantage of examining local *triangular cases* is that "most" of the inconsistencies are discovered early and can be taken into account in just the *current context* of the three concepts involved. Further, not as yet known p-conditionings can be generated and the associated probability ranges can be stepwise refined. In the general case, testing probabilistic consistency leads for every p-conditioning to a successive computing of the intersections of probability ranges derived on the basis of different local examinations.

Related probabilistic constraints have been independently examined in [Thöne *et al.*, 1992] and [Dubois *et al.*, 1992]. Note that the terminological formalism of $\mathcal{ALCP}$ allows for subsumption computation and for correctly handling logically interrelated concepts. As consequence the integrated terminological and probabilistic formalism is able to apply *refined* constraints if necessary [Heinsohn, 1991].

6 Conclusions

We have presented the language $\mathcal{ALCP}$ which is a probabilistic extension of terminological logics. The knowledge that $\mathcal{ALCP}$ allows us to handle includes *terminological knowledge* covering term descriptions and *uncertain knowledge* about (not generally true) concept properties. For this purpose, the notion of *probabilistic conditioning* based on a statistical interpretation has been introduced. The developed formal framework for terminological and probabilistic language constructs has been based on *classes of probabilities* that offer a modeling of *ignorance* as one special feature. *Probabilistic constraints* allow the context-related generation and refinement of p-conditionings and check the consistency of the knowledge base. The results of the constraints essentially depend on the correctness of the terminology which is guaranteed by the subsumption algorithm. More details about $\mathcal{ALCP}$, its formal framework, the associated interval-valued constraints, proofs, and related work can be found in [Heinsohn, 1993]. There, an assertional formalism for handling *individuals* and associated beliefs is also described.

References

[Brachman and Schmolze, 1985] R. J. Brachman and J. G. Schmolze. An overview of the KL-ONE knowledge representation system. *Cognitive Science*, 9(2):171–216, 1985.

[Dubois *et al.*, 1992] D. Dubois, H. Prade, L. Godo, and R. L. de Mantaras. A symbolic approach to reasoning with linguistic quantifiers. In *Proceedings of the 8th Conference on Uncertainty in Artificial Intelligence*, Stanford, Cal., July 1992.

[Heinsohn *et al.*, 1994] J. Heinsohn, D. Kudenko, B. Nebel, and H.-J. Profitlich. An empirical analysis of terminological representation systems. *Artificial Intelligence Journal*, 1994. Accepted for publication. A preliminary version is available as DFKI Research Report RR-92-16.

[Heinsohn, 1991] J. Heinsohn. A hybrid approach for modeling uncertainty in terminological logics. In R. Kruse and P. Siegel, editors, *Symbolic and Quantitative Approaches to Uncertainty*, LNCS 548, pages 198–205. Springer, Berlin, Germany, 1991.

[Heinsohn, 1993] J. Heinsohn. *ALCP – Ein hybrider Ansatz zur Modellierung von Unsicherheit in terminologischen Logiken*. PhD thesis, Universität des Saarlandes, Saarbrücken, 1993.

[Kruse *et al.*, 1991] R. Kruse, E. Schwecke, and J. Heinsohn. *Uncertainty and Vagueness in Knowledge Based Systems: Numerical Methods*. Series Artificial Intelligence. Springer, Berlin, Germany, 1991.

[Nebel, 1990] B. Nebel. *Reasoning and Revision in Hybrid Representation Systems*. LNAI 422. Springer, Berlin, Germany, 1990.

[Patil *et al.*, 1992] R. S. Patil, R. E. Fikes, P. F. Patel-Schneider, D. McKay, T. Finin, T. Gruber, and R. Neches. The DARPA knowledge sharing effort: Progress report. In B. Nebel, W. Swartout, and

C. Rich, editors, *Principles of Knowledge Representation and Reasoning: Proceedings KR'92*, pp. 777–788, Morgan Kaufmann, 1992.

[Schmidt-Schauß and Smolka, 1991] M. Schmidt-Schauß and G. Smolka. Attributive concept descriptions with complements. *Artificial Intelligence Journal*, 48(1), 1991.

[Sigart Bulletin, 1991] *Special Issue on Implemented Knowledge Representation and Reasoning Systems*, vol. 2(3). ACM Press, June 1991.

[Thöne *et al.*, 1992] H. Thöne, U. Güntzer, and W. Kießling. Towards precision of probabilistic bounds propagation. In *Proceedings of the 8th Conference on Uncertainty in Artificial Intelligence*, Stanford, 1992.

5

Theory of Fuzzy Systems

5.1. Theory of Fuzzy Systems: An Overview

Siegfried Gottwald

Abstract

As with any simple and fruitful mathematical notion also with the notion of fuzzy set a huge amount of theoretical considerations is and can be connected. These theoretical considerations concern basic, foundational aspects of that notions of fuzzy set as well as theoretical problems of specialized fields of applications of fuzzy sets and also the use of fuzzy sets instead of the usual, crisp sets e.g. in mathematical theories etc. This is, besides all applications of fuzzy sets and fuzzy methods, a large field of topics. In any case a field too large to be covered in some more specialized conference (or book). Hence not the whole area of main theoretical research in the fuzzy field can be covered here, and only a few remarks shall be devoted to these topics.

1 Foundational Aspects

Already from the very beginning the theory of fuzzy sets was seen under two different aspects: from the point of view of formal logic – and from the point of view of category theory.

1.1 Fuzzy sets – the logical point of view

The membership degree $\mu_A(a)$ to which a point $a \in \mathcal{X}$ of some universe of discourse $\mathcal{X}$ *belongs* to a fuzzy set A is used to *graduate* membership between complete, full, true ...membership, which is represented by $\mu_A(a) = 1$, and complete, definite, true ...non-membership, which is represented by $\mu_A(a) = 0$. Therefore, the membership degrees can be understood as *degrees of truth* of some statement of the form "a belongs to A". To distinguish the graded membership of fuzzy sets and the usual, crisp membership predicate $\in$, the former graded membership predicate often is written ε. Thus "a belongs to A" becomes formalized by the formula "$a \,\varepsilon\, A$".

Now as usually in using formalized languages, one has to be careful in distinguishing syntactic and semantic aspects w.r.t. such a formalized language. Fortunately, on the syntactic side, besides the introduction of this binary generalized membership predicate ε it is not necessary to fix too many details of that language. At first it suffices to take lower case letters to denote elements of the universe of discourse $\mathcal{X}$, and upper case letters for fuzzy subsets of $\mathcal{X}$, i.e. for elements of $\mathbb{F}(\mathcal{X})$. On the semantic side one basically needs to have a truth degree connected with each well-formed formula of the language. Denoting for any formula H this truth degree with $[\![H]\!]$, of course one has to define basically for all $a \in \mathcal{X}$ and all $A \in \mathbb{F}(\mathcal{X})$:

$$[\![a \,\varepsilon\, A]\!] =_{def} \mu_A(a). \tag{1}$$

The set of truth degrees, of course, is just the set of membership degrees, thus usually the real interval [0,1]. But this means that one immediately is in the realm of many-valued logic.

Therefore now – syntactically as well as semantically – one has to decide about the choice of logical operators, on the sentential as well as on the quantificational level. Usually there is a standard choice of the quantifiers: one has generalization $\forall$ and existential quantification $\exists$, and one interprets $\forall$ via the infimum and $\exists$ via the supremum of the respective truth degrees (of the matrix that follows the quantifier string $\forall x$ or $\exists x$). Only recently [Thiele 1993] has seriously discussed a wider class of quantifiers.

Postponing the considerations of the sentential connections for a moment, what one also needs is a flexible notation for fuzzy sets, preferably something like the usual class term notation for sets. And indeed, this notation is easily generalized. Supposing that a fuzzy set $A \in \mathbb{F}(\mathcal{X})$ is described by a well-formed formula $H(x)$ of our formalized language of fuzzy set theory, i.e. supposing that one has $\mu_A(a) = [\![H(a)]\!]$ for all $a \in \mathcal{X}$, then this fuzzy set A shall also be denoted $\{x \,\|\, H(x)\}$. That means to define

$$A = \{x \,\|\, H(x)\} \quad \Leftrightarrow_{def} \quad \mu_A(a) = [\![H(a)]\!] \quad \text{for all } a \in \mathcal{X}. \tag{2}$$

Now one has reached a notational basis which is flexible and useful, and which allows to develop large parts of fuzzy set theory in quite strong analogy with usual set theory; cf. [Bandemer, Gottwald 1993, Gottwald 1993].

To give an example, we write down the usual, minimum and maximum based intersection $A \cap B$ and union $A \cup B$ of fuzzy sets $A, B \in \mathbb{F}(\mathcal{X})$, denoting min by $\wedge$ and max by $\vee$:

$$A \cap B = \{x \,\|\, x \,\varepsilon\, A \wedge x \,\varepsilon\, B\}, \qquad (3)$$

$$A \cup B = \{x \,\|\, x \,\varepsilon\, A \vee x \,\varepsilon\, B\}. \qquad (4)$$

1.2 Fuzzy logic as the logic for fuzzy sets theory

Up to now we did not discuss the sentential connectives to be used in the many-valued logic for fuzzy sets. Surely, a conjunction $\wedge$ and a disjunction $\vee$ with truth degree functions min, max respectively should appear. Additionally, connected with the complement $\overline{A}$ of $A \in \mathbb{F}(\mathcal{X})$, defined by: $\mu_{\overline{A}}(a) = 1 - \mu_A(a)$, one considers a negation connective $\neg$ with truth degree function $1 - \ldots$, such that one has $\overline{A} = \{x \,\|\, \neg(x \,\varepsilon\, A)\}$.

But, contrary to the situation in classical logic, this set $\wedge, \vee, \neg$ of connectives is not functionally complete; cf. [Gottwald 1989]. Furthermore, already in [Zadeh 1965] additional intersections and unions besides (3), (4) have been introduced: the (algebraic) product which can be considered as a variant of (3) based on another conjunction connective with the product as its truth degree function, and the (bounded) sum which is a variant of (4) based on another disjunction connective with $\min\{1, u+v\}$ as its truth degree function.

Of course, a lot more such versions can be constructed. What one is looking for, hence, is some framework for a more general discussion of either the set algebra for fuzzy sets or the many-valued sentential logic which it is based upon. We here will look at the sentential logic. What we thus have to look for are either some uniform ways to introduce further sentential connectives besides $\wedge, \vee, \neg$ or to enrich the complete distributive lattice $\langle [0,1], \wedge, \vee \rangle$ with additional operations.

Both these possibilities are under discussion and even parallel one another to a large extend. Actually, the prefered way to introduce new sentential connectives is to start from new types of conjunction connectives $\&_{\boldsymbol{t}}$ based on a t-norm $\boldsymbol{t}$; cf. [Gottwald 1993, Gupta, Qi 1991]. A t-norm $\boldsymbol{t}$ is a binary operation in [0,1] which is commutative, associative, monotonously nondecreasing in both arguments, and which has 1 as a unit and 0 as a zero element. The class of all t-norms is quite large, therefore subclasses or even (one-)parametric families of t-norms have been introduced and sometimes are used only.

With each t-norm $\boldsymbol{t}$ via a de Morgan connection a unique t-conorm

$\boldsymbol{s_t}$ is joined as $\boldsymbol{s_t}(u,v) = 1 - \boldsymbol{t}(1-u, 1-v)$. Each such t-conorm $\boldsymbol{s_t}$ is the truth degree function of a disjunction connective $\vee_{\boldsymbol{t}}$, and the definition of $\boldsymbol{s_t}$ forces that $[\![H_1 \vee_{\boldsymbol{t}} H_2]\!] = [\![\neg(\neg H_1 \,\&_{\boldsymbol{t}}\, H_2)]\!]$ holds true for all well-formed formulas H_1, H_2 of the formalized language of fuzzy set theory. Furthermore, with each left continuous t-norm $\boldsymbol{t}$ in a unique way, analoguous to the relative pseudo-complementation in lattice theory (cf. [Rasiowa 1974]), i.e. analoguous to Heyting algebras, an often so-called Φ-operator $\varphi_{\boldsymbol{t}}$ is connected as a residuation by

$$\varphi_{\boldsymbol{t}}(u,v) = \sup\{w \mid \boldsymbol{t}(u,w) \leq v\} \tag{5}$$

such that one always has true the characteristic condition

$$\boldsymbol{t}(u,w) \leq v \quad \Leftrightarrow \quad w \leq \varphi_{\boldsymbol{t}}(u,v). \tag{6}$$

These Φ-operators $\varphi_{\boldsymbol{t}}$ in a natural way are truth degree functions of implication connectives $\rightarrow_{\boldsymbol{t}}$; cf. [Gottwald 1993].

By the way it is interesting to recognize that the left continuity of $\boldsymbol{t}$ just means that one always has

$$[\![\exists x(H(x) \,\&_{\boldsymbol{t}}\, G)]\!] = [\![\exists x H(x) \,\&_{\boldsymbol{t}}\, G]\!] \tag{7}$$

for any well-formed formulas $H(x), G$ such that G does not contain the variable x free.

Thus, starting from a given t-norm $\boldsymbol{t}$, one has always a whole family $\&_{\boldsymbol{t}}, \vee_{\boldsymbol{t}}, \rightarrow_{\boldsymbol{t}}$ of sentential connectives, usually besides $\wedge, \vee$. But even $\wedge, \vee$ fit into this framework as $\wedge = \&_{\mathsf{min}}, \vee = \vee_{\mathsf{max}}$. And thus one has found a sufficiently rich formal language for fuzzy set theory. Without further set algebraic details we simply illustrate the expressive power with the definition of a graded, t-norm based inclusion relation $\subseteq_{\boldsymbol{t}}$ as

$$A \subseteq_{\boldsymbol{t}} B \Leftrightarrow_{def} \forall x(x \,\varepsilon\, A \rightarrow_{\boldsymbol{t}} x \,\varepsilon\, B). \tag{8}$$

On the present basis then the way is open for deeper investigations into fuzzy set theory and its logic; cf. [Gottwald 1993, Takeuti, Titani 1992].

1.3 Algebraic structures for truth/membership degrees

Algebraically, the discussions of the preceding section are concerned with the structure of the set of truth/membership degrees [0,1]. With $\wedge =$ min and $\vee =$ max this set becomes a complete distributive lattice. An algebraic problem now is if this lattice structure here is the right one, or if some other, enriched structure should have preference. Indeed, the class $\mathbb{F}(\mathcal{X})$ of all fuzzy subsets of the universe of discourse $\mathcal{X}$ with the operations $\cap, \cup$ is just the direct product $\prod_{x \in \mathcal{X}} \mathcal{I}$ of card$(\mathcal{X})$ copies of the lattice $\mathcal{I} = \langle [0,1], \wedge, \vee \rangle$. And what is known as L-fuzzy sets is just the result of an exchange of this lattice $\mathcal{I}$ for another lattice L in this construction.

The introduction of t-norms, yet, does not correspond to this idea of L-fuzzy sets: $\langle [0,1], \boldsymbol{t}, \boldsymbol{s_t} \rangle$ is a lattice only for $\boldsymbol{t} = \min$. This idea of t-norm introduction instead corresponds to the introduction of an enriched structure into the lattice $\mathcal{I}$. The algebraic problem, but, is which type of enriched structure is the appropriate one.

There is general agreement that one of the essential properties of all the t-norms $\boldsymbol{t} \neq \min$ is that they are not idempotent, i.e. that for $\boldsymbol{t} \neq \min$ there always exist $u \in [0,1]$ with $\boldsymbol{t}(u,u) < u$. Thus, only enriching the lattice $\mathcal{I}$ with the negation $\neg$ and hence studying *de Morgan algebras* is not sufficient. Even the fact that via residuation (5), (6) also with $\wedge = \min$ an implication operator $\rightarrow_G$, the well known Gödel implication of many-valued logic, can be connected – or algebraically: that one can go from the lattice $\mathcal{I}$ to some *Heyting algebra* does not really solve the problem to enrich the lattice structure by a non-idempotent "product". Therefore one algebraically prefers to change from the lattice $\mathcal{I}$ of truth/membership degrees either to a residuated lattice in the sense of [Dilworth, Ward 1938] or to an MV-algebra in the sense of [Chang 1958]; cf. e.g. [Höhle, Stout 1991, Turunen 1992]. In any case, once again, the main point of this enrichment is to have for the logic a canonical implication operator and together with the lattice meet a further non-idempotent, i.e. "*interactive*" product.

This non-idempotent additional product, which is related to a non-idempotent conjunction connective, seems to be highly characteristic for logico-algebraic considerations connected with fuzzy sets. And having such a non-idempotent product $*$ creates the problem of its invertibility. One aspect of this problem is to look for any given element v of the algebraic structure for the solvability of the equation $u * u = v$. Such a solution can be seen as a $*$-root of v. And just this notion of root, the problem of the existence of such roots, and consequences of the existence of such roots is the topic of the paper of HÖHLE which follows in this chapter.

1.4 Fuzzy sets – the categorical point of view

The more abstract point of view of category theory adds to the foregoing discussions on the structure of the truth/membership degrees, i.e. to the set algebraic structure of $\mathbb{F}(\mathcal{X})$ the considerations of morphisms between fuzzy sets. The objects, of course, of the categories of fuzzy sets to be introduced are the fuzzy sets.

Quite early, within the first decade [Goguen 1974] (with a few remarks already in [Goguen 1967]) has given a categorical framework for fuzzy sets. But for long this was seen as a more or less esoteric addition to the mainstream discussions in the fuzzy field. Work by Manes, e.g. [Arbib, Manes 1975, Manes 1982], later offered the idea that graded membership together with graded equalitiy should be of basic importance for fuzzy sets, and that there should be connections between categories for fuzzy sets and toposes.

This gave an impulse which in recent years much revived the categorical investigations into fuzzy sets. For the details we refer to the literature, e.g. [Höhle, Stout 1991, Stout 1991, Rodabaugh et al. 1992]. At present the most interesting outcome of these works seem to be that they support the view that a fuzzy set always is strongly tied with a graded, i.e. many-valued equality relation in its universe of discourse. And this point of view even becomes essential for approaching fuzzy control; cf. [Kruse et al. 1993].

2 Fuzzy Control and Fuzzy Relation Equations

Actually fuzzy control is the most prominent applicational area for the fuzzy sets idea. Of course, a lot of the problems in fuzzy control are of engineering nature and related to each specific application. But there are also common features for all the fuzzy control applications, viz. general aspects of the fuzzy control methodology. And, moreover, some such general aspects are capable of a uniform theoretical treatment.

The general starting point for the theoretical understanding of fuzzy control is that such a control algorithm is presented in the form of a finite list of *control rules* which in the simplest case have the form

$$\text{IF} \quad \alpha = A_i \quad \text{THEN} \quad \beta = B_i, \qquad i = 1, \ldots, n \tag{9}$$

with α the input variable and β the output variable of the control algorithm. Here all the A_i, B_i are fuzzy subsets of suitable universes of discourse $\mathcal{X}, \mathcal{Y}$, respectively.

Essentially there are two variants to approach the mathematical realization of such a list (9) of control rules. The pioneering approach of [Mamdani, Assilian 1975] used the transformation of (9) into a fuzzy relation R and the compositional rule of inference of [Zadeh 1973] to connect with each fuzzy input value $\alpha = A$ a fuzzy output value $\beta = B$ as

$$B = A \circ R = \{y \,\|\, \exists x (x \,\varepsilon\, A \wedge (x,y) \,\varepsilon\, R)\}. \tag{10}$$

The first industrial application [Holmblad, Østergaard 1982] to the cement kiln process connected with each fuzzy input $\alpha = A$ and each control rule of (9) a *degree of activation*

$$\gamma_i(A) = \mathrm{hgt}\,(A_i \cap A) = [\![\exists x (x \,\varepsilon\, A_i \wedge x \,\varepsilon\, A)]\!] \tag{11}$$

and reached an output $\beta = B$ by superposition of the rules outputs as

$$B = \bigcup_{i=1}^{n} \gamma_i \cdot B_i = \bigcup_{i=1}^{n} \{y \,\|\, (y \,\varepsilon\, B_i) \,\&_{\boldsymbol{t}_1}\, \gamma_i(A)\} \tag{12}$$

with the t-norm $\boldsymbol{t}_1 =$ product. A slight reformulation of (12) gives

$$B = \bigcup_{i=1}^{n} \{y \,\|\, \exists x (x \,\varepsilon\, A_i \wedge x \,\varepsilon\, A) \,\&_{\boldsymbol{t}_1}\, y \,\varepsilon\, B_i\}. \tag{13}$$

Thus (13) is not too different of (10). The analogy of both approaches becomes even more obvious if one looks at the special choice of the fuzzy relation R_0 in [Mamdani, Assilian 1975]: $R_0 = \bigcup_{i=1}^{n}(A_i \times B_i)$ with $A_i \times B_i = \{(x,y) \,\|\, x \,\varepsilon\, A_i \wedge x \,\varepsilon\, B_i\}$. With this specific relation R_0 formula (10) becomes, cf. [Gottwald 1993]:

$$B = \bigcup_{i=1}^{n} \{y \,\|\, \exists x (x \,\varepsilon\, A_i \wedge x \,\varepsilon\, A) \wedge y \,\varepsilon\, B_i\}. \tag{14}$$

The implementation of both these fuzzy control strategies is quite straightforward. Hence they give a basis for automatic control devices. Even crisp inputs are easy to handle: they are understood as fuzzy singletons. But the output usually is a fuzzy set. And then to extract from it – e.g. automatically – a final control action to be taken, this means that one has to *defuzzify* this fuzzy output. For this defuzzification process one has some proposals how to proceed, but essentially no theory up to now. The following paper of RUNKLER/GLESNER is an partial attempt to reach such a theory.

The idea to realize a fuzzy controller via (10) by a fuzzy relation can also be seen from another point of view. Having (9) realized by R would mean to have as a special case of (10)

$$B_i = A_i \circ R, \qquad i = 1, \ldots, n. \tag{15}$$

Unfortunately, only under additional assumptions concerning the input/output data A_i, B_i equations (15) are satisfied by Mamdani's relation R_0. Usually, the control rules (9) will *interact*; cf. [Gottwald 1993]. To avoid this interactivity effect one can turn the problem and take (15) as a system of relation equations which has to be solved for the unknown fuzzy relation R. This idea transforms the determination of a fuzzy relation R out of the control rules (9) into the mathematical problem to solve system (15).

There is a lot of theory concerning solvability and approximate solvability of such fuzzy relation equations and systems of them which shall not be reviewed here; cf. e.g. [diNola et al. 1989, Gottwald 1993].

But the comparison of the two control strategies mentioned above, i.e. of (14) and (13) shows also that the compositional rule of inference (10) is not the only way to get a fuzzy output from a fuzzy input and a fuzzy (control) relation. Mathematically this means that also other types of relation equations besides (15) become of interest. And the following paper of GOTTWALD just concerns the extension of the solvability theory of fuzzy relation equations to other types of equations.

Fuzzy control of course competes with traditional control methods. Therefore it is of theoretical interest to know about the types of control functions which one is able to realize by fuzzy controllers. More recent work like [Buckley 1993, Buckley, Hayashi 1993] treats this problem. But there is another aspect of the same problem: to try to construct a fuzzy controller which realizes a given control function. If one is able to do this in a somewhat uniform manner, then a way is opened for some type of mathematical comparison of fuzzy and traditional control. The following paper of BAUER et al. is concerned with this problem.

3 Measuring Fuzziness and Fuzzy Information

Intuitively the most essential type of usage of fuzzy sets is to have them available for coding vague as well as imprecise, "qualitative" informations. This e.g. is the main reason behind fuzzy and linguistic variables, and also the main background for the whole area of approximate reasoning.

Of central importance for the treatment of such qualitative informations are all methods which are concerned with the inference of "new", further informations out of given ones. With these problems one is concerned e.g. in discussions which combine fuzzy set tools with developments in Artificial Intelligence, in Expert Systems, in Neural Nets etc., of course also in fuzzy logic – this last term now taken for systems of formalized logic which allow the treatment of fuzzy sets of well-formed formulas as sets of axioms or of premisses; cf. e.g. [Kruse et al. 1991, Novak 1989].

Besides this intuitive notion of (qualitative) information traditional information theory has a numerical measure for information (contents). Its central numerical notion is that one of entropy.

For fuzzy sets, on the other hand, one has to look for an additional aspect here: it is not only "information content" or "capacity for information transmittal" one may be interested in, it is also the question of "how fuzzy" or "how imprecise" some qualitative information is. It was already in [deLuca, Termini 1972] that this problem was taken into consideration; cf. also [deLuca, Termini 1979]. Since one has discussed such "measures of fuzziness" mainly under two aspects: (i) as *entropy measures* which characterize the deviation of a fuzzy set from a crip one, and (ii) as *energy measures* which essentially characterize the deviation of a fuzzy set from a crisp singleton; cf. [Bandemer, Gottwald 1993]. This attitude toward measures of fuzziness and entropy measures for fuzzy sets puts them into a quite narrow understanding. The following paper of SANDER intends to widen this understanding in that it embeds such entropy measure into the wider area of general entropy functions discussed in modern information theory. It remains to be seen if such an approach may open further doors for a quantitative information theory for fuzzy informations besides e.g. possibilistic information theory and Dempster-Shafer theory; cf. [Klir, Folger 1988, Klir 1991].

4 Fuzzified Mathematical Notions and Structures

Large parts of modern mathematics are seen as to be based on a set theoretic foundation, especially in topology and algebra. Thus a generalization of the notion of set can almost straightforward be used to generalize such parts of mathematics. But as always inside mathematics, such generalizations may (i) prove to be of real value and open new and fruitful areas for research, or (ii) may be of a superficial nature and nothing more than mere intellectual toying. Unfortunately, at the very beginning it is quite hard to give a serious evaluation concerning these types of generalization w.r.t. some specific topic.

In the first years after the seminal paper [Zadeh 1965] the overwhelming majority of mathematicians looked at fuzzifications of mathematical structures as at generalizations of type (ii). The only exception seemed to be the study of fuzzy relations because of their clear relationships to problems in theoretical computer science and the treatment of uncertain informations. Meanwhile, the situation has changed. The forerunner of this development inside mathematics has been fuzzy topology with its categorical methods and its applications also outside fuzzified mathematical structures like to probabilistic topological spaces, lattice theory, locales and Stone representation theory; cf. [Lowen et al. 1991, Rodabaugh 1991, Rodabaugh et al. 1992].

Not so clear, actually, is the situation w.r.t. fuzzified algebraic structures like fuzzy groups etc. Their definitions essentially follow a standard pattern: given a crisp basic structure like a group $\mathcal{G} = \langle G, \cdot, ^{-1} \rangle$, a fuzzy subgroup is some $S \in \mathbb{F}(\mathcal{G})$ such that

$$[\![x \,\varepsilon\, S \,\&_{\boldsymbol{t}}\, y \,\varepsilon\, S \rightarrow_{\boldsymbol{t}} x \cdot y^{-1} \,\varepsilon\, S]\!] = 1 \tag{16}$$

holds true for all $x, y \in \mathcal{G}$ and some t-norm $\boldsymbol{t}$. Actually, there is a lot of nice mathematics which can be done on such a basis; cf. [Wang 1993]. Nevertheless, it is essentially an open problem which true applications outside fuzzified mathematics these structures will have. Therefore we shall not go into more details here.

References

[Arbib, Manes 1975] M.Arbib, E.G.Manes: A category theoretic approach to systems in a fuzzy world. Synthese 30 (1975), 381-406.

[Bandemer, Gottwald 1993] H.Bandemer, S.Gottwald: Einführung in Fuzzy-Methoden. Aufbau-Verlag, Berlin (1993).

[Buckley 1993] J.J.Buckley: Controllable processes and the fuzzy controller. Fuzzy Sets Syst. 53 (1993), 27-31.

[Buckley, Hayashi 1993] J.J.Buckley, Y.Hayashi: Fuzzy input-output controllers are universal approximators. Fuzzy Sets Syst. 58 (1993), 273-278.

[Chang 1958] C.C.Chang: Algebraic analysis of many-valued logics. Trans. Amer. Math. Soc. 88 (1958), 467-490.

[deLuca, Termini 1972] A.de Luca, S.Termini: A definition of non-probabalistic entropy in the setting of fuzzy sets theory. Inform. and Control 20 (1972), 301-312.

[deLuca, Termini 1979] A.deLuca, S.Termini: Entropy and energy measures of a fuzzy set. In: M.M.Gupta, R.K.Ragade, R.R.Yager (eds.): Advances in Fuzzy Set Theory and Applications. North-Holland, Amsterdam, 321-338 (1979).

[Dilworth, Ward 1938] R.P.Dilworth, M.Ward: Residuated lattices. Trans. Amer. Math. Soc. 45 (1939), 335-354.

[diNola et al. 1989] A.diNola, S.Sessa, W.Pedrycz, E.Sanchez: Fuzzy Relation Equations and Their Applications to Knowledge Engineering. Kluwer Acad. Publ., Dordrecht (1989).

[Goguen 1967] J.A.Goguen: L-fuzzy sets. J. Math. Anal. Appl. 18 (1967), 145-174.

[Goguen 1974] J.A.Goguen: Concept representation in natural and artificial languages: axioms, extensions and applications for fuzzy sets. Internat. J. Man-Machine Stud. 6 (1974), 513-561.

[Gottwald 1989] S.Gottwald: Mehrwertige Logik. Aufbau-Verlag, Berlin (1989).

[Gottwald 1993] S.Gottwald: Fuzzy Sets and Fuzzy Logic. Vieweg-Verlag, Braunschweig (1993).

[Gupta, Qi 1991] M.M.Gupta, J.Qi: Theory of t-norms and fuzzy inference methods. Fuzzy Sets Syst. 40 (1991), 431-450.

[Höhle, Stout 1991] U.Höhle, L.N.Stout: Foundations of fuzzy sets. Fuzzy Sets Syst. 40 (1991), 257-296.

[Holmblad, Østergaard 1982] L.P.Holmblad, J.J.Østergaard: Control of cement kiln by fuzzy logic. In: M.M.Gupta, E.Sanchez (eds.): Fuzzy Information and Decision Processes, North-Holland, Amsterdam, 389-399 (1982).

[Klir, Folger 1988] G.J.Klir, T.A.Folger: Fuzzy Sets, Uncertainty, and Information. Prentice Hall, Englewood Cliffs NJ (1988).

[Klir 1991] G.J.Klir: Generalized information theory. Fuzzy Sets Syst. 40 (1991), 127-142.

[Kruse et al. 1991] R.Kruse, E.Schwecke, J.Heinsohn: Uncertainty and Vagueness in Knowledge Based Systems. Springer-Verlag, Berlin (1991).

[Kruse et al. 1993] R.Kruse, J.Gebhardt, F.Klawonn: Fuzzy-Systeme. Teubner, Stuttgart (1993).

[Lowen et al. 1991] E.Lowen, R.Lowen, P.Wuyts: The categorical topology approach to fuzzy topology and fuzzy convergence. Fuzzy Sets Syst. 40 (1991), 347-373.

[Mamdani, Assilian 1975] E.H.Mamdani, S.Assilian: An experiment in linguistic synthesis with a fuzzy logic controller. Internat. J. Man-Machine Stud. 7 (1975), 1-13.

[Manes 1982] E.G.Manes: Review of "Fuzzy Sets and Systems: Theory and Applications" by Dubois& Prade. Bull. Amer. Math. Soc. 7 (1982), 603-612.

[Novak 1989] V.Novak: Fuzzy Sets and Their Applications. Adam Hilger, Bristol (1989).

[Rasiowa 1974] H.Rasiowa: An Algebraic Approach to Non-Classical Logics. North-Holland, Amsterdam (1974).

[Rodabaugh 1991] S.E.Rodabaugh: Point-set lattice-theoretic topology. Fuzzy Sets Syst. 40 (1991), 297-345.

[Rodabaugh et al. 1992] S.E.Rodabaugh, E.P.Klement, U.Höhle (eds.): Applications of Category Theory to Fuzzy Subsets. Kluwer Acad. Publ., Dordrecht (1992).

[Stout 1991] L.N.Stout: A survey of fuzzy set and topos theory. Fuzzy Sets Syst. 41 (1991), 3-14.

[Takeuti, Titani 1992] G.Takeuti, S.Titani: Fuzzy logic and fuzzy set theory. Arch. Math. Logic 32 (1992), 1-32.

[Thiele 1993] H.Thiele: On fuzzy quantifiers. Proc. 5. IFSA World Congress, Seoul 1993, vol. 1, 395-398 (1993).

[Turunen 1992] E.Turunen: Algebraic structures in fuzzy logic. Fuzzy Sets Syst. 52 (1992), 181-188.

[Wang 1993] P.P.Wang (ed.): Advances in Fuzzy Theory and Technology. Vol. 1, 1993. Bookwrights Press, Durham NC (1993).

[Zadeh 1965] L.A.Zadeh: Fuzzy sets. Inform. and Control 8 (1965), 338-353.

[Zadeh 1973] L.A.Zadeh: Outline of a new approach to the analysis of complex systems and decision processes. IEEE Trans. Systems, Man and Cybernet. SMC-3 (1973), 28-44.

5.2. Solvability Considerations for Non–Elementary Fuzzy Relational Equations

Siegfried Gottwald

Abstract

Determining a fuzzy controller is a two step process. First one has to fix the control rules which in general have the form: IF $\alpha = A_i$ THEN $\beta = B_i, i = 1, \ldots, N$ for some (perhaps multidimensional) input variable α and output variable β. And then one has to convert these control rules into a fuzzy relation R and to apply this relation via the compositional rule of inference to any fuzzy input value of α. Ideally the rule inputs A_i should transform into the rule outputs $B_i = A_i \circ R$. Thus, mathematically this second step can be seen as the problem to solve a system of fuzzy relational equation for an unknown fuzzy relation.

Besides these simple, "elementary" types of equations also other, more complicated types of equations have been discussed in fuzzy control theory. The present paper discusses some aspects of how results from the solvability theory of elementary fuzzy relational equations can be extended to discuss also more involved types of relational equations.

1 Elementary and non-elementary fuzzy relational equations

Fuzzy relational equations usually are written down in the form $A \circ R = B$ with fuzzy sets $A \in \mathbb{F}(\mathcal{X}), B \in \mathbb{F}(\mathcal{Y})$ and a fuzzy relation $R \in \mathbb{F}(\mathcal{X} \times \mathcal{Y})$. Here $\mathcal{X}$ and $\mathcal{Y}$ are the corresponding universes of discourse. In terms of the membership degrees this equation reads:

$$\mu_B(y) = \sup_{x \in \mathcal{X}} \min\{\mu_A(x), \mu_R(x, y)\}, \tag{1}$$

or in the terminology which explicitely refers to the background from many-valued logic:

$$[\![y \varepsilon B]\!] = [\![\exists x(x \varepsilon A \wedge (x,y) \varepsilon R)]\!]. \tag{2}$$

Here (1) is just the well known compositional rule of inference which was introduced in [Zadeh 1973].

These *elementary* fuzzy relational equations have been generalized in different ways, usually with a background in ideas related to fuzzy control.

Main types of non-elementary fuzzy relational equations studied so far and which are of special significance for applications in the field of fuzzy modelling include the following:

(1) fuzzy relational equations $B = A \circ_t R$ with sup-t-composition

$$\mu_B(y) = \sup_{x \in \mathcal{X}} (\mu_A(x) \, t \, \mu_R(x,y)), \tag{3}$$

which is strongly connected with the compositional rule of inference.

(2) fuzzy relational equations $B = A \diamond_t R$ with inf-s-composition

$$\mu_B(y) = \inf_{x \in \mathcal{X}} (\mu_A(x) \, s_t \, \mu_R(x,y)), \tag{4}$$

which in a suitable sense are dual to the foregoing case;

(3) adjoint fuzzy relational equations $B = A \nabla_t R$ of [Pedrycz 1985] with inf-φ-composition

$$\mu_B(y) = \inf_{x \in \mathcal{X}} (\mu_A(x) \, \varphi_t \, \mu_R(x,y)). \tag{5}$$

Besides these types, which might be viewed as being of a basic nature, one can consider some types of a more complex form which sometimes are formed on the basis of the types given above. Of such possibilities two shall be mentioned:

(4) a convex combination $B = \lambda \cdot (A \circ_t R_1) + (1 - \lambda) \cdot (A \diamond_{s_{t_1}} R_2)$ of equations of types (3) and (4) as in [Ohsato, Sekiguchi 1983]

$$\begin{aligned} \mu_B(y) = \lambda(y) \cdot (&\sup_{x \in \mathcal{X}} (\mu_A(x) \, s_t \, \mu_{R_1}(x,y))) + \\ &+ (1 - \lambda(y)) \cdot \inf_{x \in \mathcal{X}} (\mu_A(x) \, s_{t_1} \, \mu_{R_2}(x,y)) \end{aligned} \tag{6}$$

with $\lambda : \mathcal{Y} \to [0,1]$ and with the t-norm t, as well as the t-conorm s_{t_1}, in [Ohsato, Sekiguchi 1983] taken as max, min only;

(5) fuzzy relational equations $B = A \bowtie_t R$ with an equality operator as discussed in [diNola et al. 1988]

$$\mu_B(y) = \sup_{x \in \mathcal{X}} (\mu_A(x) \varphi_t \mu_R(x,y) \wedge (\mu_R(x,y) \varphi_t \mu_A(x))). \tag{7}$$

For all the basic forms of equations (3) – (5) the family of solutions, if solutions do exist at all, has been characterized and their extremal (maximal or minimal) elements have been obtained. Moreover, for systems

$$B_i = A_i \circ_t R, \quad i = 1, \ldots, N \tag{8}$$
$$B_i = A_i \diamond_t R, \quad i = 1, \ldots, N \tag{9}$$
$$B_i = A_i \nabla_t R, \quad i = 1, \ldots, N \tag{10}$$

of such equations the relevant results are also available – mainly under the condition that they do have an exact solution.

To have an overall picture of the results they are collected in Table 1.

Type of equation	Solution[1] to a single equation	Solution[1] to a system of equations
$B = A \circ_t R$, sup-t-composition	$\hat{R} = A \rhd_t B$ $\hat{R} = \sup \boldsymbol{R}'$	$\hat{R} = \bigcap_{i=1}^{n} (A_i \rhd_t B_i)$ $\hat{R} = \bigcap_{i=1}^{n} \sup \boldsymbol{R}'_i$
$B = A \diamond_t R$, inf-s-composition	$\check{R} = A \lhd_t B$ $\check{R} = \inf \boldsymbol{R}''$	$\check{R} = \bigcup_{i=1}^{n} (A_i \lhd_t B_i)$ $\check{R} = \bigcup_{i=1}^{n} \inf \boldsymbol{R}''_i$
$B = A \rhd_t R$, inf-φ-composition	$\tilde{R} = A \times_t B$ $\tilde{R} = \inf \boldsymbol{R}'''$	$\tilde{R} = \bigcup_{i=1}^{n} (A_i \times_t B_i)$ $\tilde{R} = \bigcup_{i=1}^{n} \inf \boldsymbol{R}'''_i$

Table 1: Basic non-elementary types of fuzzy relation equations for control applications

[1] in case of solvability

Concerning the notation used in this table we have to add some explanations. First, by $\boldsymbol{R}', \boldsymbol{R}'', \boldsymbol{R}'''$ we denote the sets of solutions of the systems (8), (9), (10) respectively of fuzzy relational equations. Secondly, related with the Φ-operator φ_t one has a kind of "product" $\rhd_t$ giving a fuzzy relation $T := A \rhd_t B$ out of fuzzy sets A, B defined by

$$\mu_T(x, y) =_{def} \mu_A(x) \varphi_t \mu_B(y). \tag{11}$$

And finally as a dual to the Φ-operator φ_t we use the β-operator defined for all $u, v \in [0, 1]$ as:

$$u \beta_t v =_{def} \inf\{w \in [0, 1] \mid u s_t w \geq v\} \tag{12}$$

and connect with it now an operator $\lhd_t$ "dual" to $\rhd_t$ which for fuzzy sets A, B again gives a fuzzy relation $S := A \lhd_t B$ characterized by

$$\mu_S(x, y) =_{def} \mu_A(x) \beta_t \mu_B(y). \tag{13}$$

At present however, for the general case it seems quite difficult to give simple and easy-to-check conditions for the solvability of a system of equations. This was one of the reasons for the discussion of degrees of solvability for example in [Gottwald 1986]. Thus, the results e.g. of Table 1 have a significant value only in the case that solutions really exist, i.e. that not only "approximate" solutions (in some suitable sense of that word) exist. If this true solvability is not the case – and it is this more uncomfortable situation one usually meets in practice – then because of the (present) lack of an extended mathematical theory of such (systems of) equations, the user has to think about other ways of overcoming the problem of the nonexistence of (true) solutions.

A simple, and perhaps for the practitioner the most obvious, way out is to use the formulas which describe solutions – in the case of solvability – even if a solution to the system of fuzzy relation equations to be considered does not exist – and then to check the quality of the "approximate solution" derived in this way.

But having taken this point of view one can move one step further: instead of having proven a formula to give a solution in the case of solvability, one can start from a formula which one guesses to describe a solution – of course, if there are some acceptable reasons for such a guess. And, indeed, for some classes of fuzzy relational equations such acceptable guesses are available. To present some basic ones let us distinguish for relational equations $\Theta(R, A) = B$ two different types.

Definition 1 *A fuzzy equation $\Theta(R, A) = B$ will be said to be of* sup-type *in the case that one has*

$$\mu_B(y) = \mu_\Theta(A, R)(y) = \sup_{x \in \mathcal{X}} \Gamma(\mu_A(x), \mu_R(x, y)) \tag{14}$$

where the term Γ is built up using the membership degrees $\mu_A(a)$, $\mu_R(a, b)$ and combining them for example by a t-norm, a Φ-operator or some suitable other kinds of "simple" operators; and such an equation will be said to be of inf-type *in the case that one has accordingly*

$$\mu_B(y) = \mu_\Theta(A, R)(y) = \inf_{x \in \mathcal{X}} \Gamma(\mu_A(x), \mu_R(x, y)). \tag{15}$$

To show the influence of this distinction on the structure of the set of solutions of these equations we will consider the following facts which should be taken into account for the discussions of true and of approximate solutions.

Fact 1: The union of any two solutions of a fuzzy relational equation of sup-type is again a solution to this equation, and hence this equation has a greatest solution.

Fact 2: The intersection of any two solutions of a fuzzy relational equation of inf-type is again a solution to this equation, and hence this equation has a smallest solution.

Fact 3: Any system of fuzzy relational equations of sup-type has in the case of solvability as the greatest solution the intersection of all the greatest solutions of its single equations.

Fact 4: Any system of fuzzy relational equations of inf-type has in the case of solvability as the smallest solution the union of all the smallest solutions of its single equations.

It is interesting to mention that these types of behaviour can also be found with some mixed types of fuzzy relational equations. To look at an example we refer to the convex combination form (6), (6) of fuzzy relational equations discussed by [Ohsato, Sekiguchi 1983, 1985] for $A \circ_t R_1$ as sup-min-composition and $A \diamond_t R_2$ as inf-max-composition. Indeed, in [Ohsato, Sekiguchi 1985] a solvability behaviour of such equations is proven which combines Facts 1 and 2: the type (6) of equations has an extremal solution $(\check{R}_1, \hat{R}_2)$ in the sense that $R'_1 \subseteq \check{R}_1$ and $R'_2 \subseteq \hat{R}_2$ for all solutions (R'_1, R'_2) to this equation. Extending these results one can prove that for solutions (R'_1, R'_2) and (R''_1, R''_2) of this equation also $(R'_1 \cup R''_1, R'_2 \cap R''_2)$ is a solution. And extending Facts 3 and 4 one can prove that an extremal solution of a system of such equations (6) is determined – in the case of its existence – in its "first coordinate" as the intersection of the "first coordinates" of the extremal solutions of the single equations, and in its "second coordinate" as the union of the "second coordinates" of the extremal solutions of the single equations.

2 Coupled pairs of non-elementary fuzzy relational equations

Having designed a system of control rules for a fuzzy controller means that one has given a family of associated fuzzy sets

$$(A_1, B_1), (A_2, B_2), \ldots, (A_N, B_N). \tag{16}$$

From a very general point of view now one has to determine a "relational structure" $\mathcal{R}$ such that the fuzzy sets (16) satisfy it, i.e. that one always has $\mathcal{R}(A_k, B_k)$. Caused by the fact that such a relational structure $\mathcal{R}$ has an internal (logical) structure one thus has to reveal some logical structure within the data.

The problem given in such a general setting requires further clarification, in particular with respect to the structure of $\mathcal{R}$. Furthermore here our consideration is influenced by some ideas which relate fuzzy control and neural net techniques. Generally now the "relational structure" shall be treated as a cascade of fuzzy relational equations. These equations involve fuzzy sets and fuzzy relations combined by the relational composition operators $\circ$ for max-t composition and $\#$ for min-s composition, t any t-norm and s their t-conorm. From the point of view of fuzzy relational equations we therefore are concerned with equations of the types

$$B = A \circ R \qquad \text{and} \qquad B = A \# R. \tag{17}$$

The relational structure $\mathcal{R}$ with its cascade structure shall be described by nesting of those simple types of relational equations. For simplicity we will restrict ourselves here to a two-level cascade structure with different standard compositions of both levels. This two-level structure implies that we have to consider an intermediate layer besides the input and output layers. The general construction of an intermediate layer between input and output fuzzy sets now means to consider e.g. coupled systems of fuzzy relational equations of the kind

$$B = Z \# G \quad \text{and} \quad Z = A \circ R, \tag{18}$$

or of the kind

$$B = Z \circ G \quad \text{and} \quad Z = A \# R. \tag{19}$$

In more explicite notation that means e.g. for the second case that one has

$$\mu_B(y) = \sup_{z \in \mathcal{Z}} [\mu_Z(z) \, \boldsymbol{t} \, \mu_G(z, y)], \qquad \mu_Z(z) = \inf_{x \in \mathcal{X}} [\mu_A(x) \, \boldsymbol{s} \, \mu_R(x, z)]$$

with an additional universe $\mathcal{Z} = \{z_1, z_2, \ldots, z_p\}$ for the intermediate fuzzy sets. The essential role of that additional universe is to enhance the representation capabilities of the input/output mapping via our cascading relational structure $\mathcal{R}$.

The original problem can be seen as that one of solving systems of relational equations originating out of such a structure.

Bearing in mind the interpretation of t-norms as AND connectives and of t-conorms as OR connectives the system $\mathcal{R}$ conveys a clear structural interpretation. The min-$\boldsymbol{s}$ composition performs the role of an AND neuron combining inputs $A(x_1), A(x_2), \ldots, A(x_n)$ – concisely

$$Z(z_l) = \text{AND}(A, R_l)$$

or coordinatewise

$$Z(z_l) = [A(x_1) \,\text{OR}\, R(x_1, z_l)] \,\text{AND}\, [A(x_2) \,\text{OR}\, R(x_2, z_l)] \,\text{AND} \\ \ldots \,\text{AND}\, [A(x_n) \,\text{OR}\, R(x_n, z_l)].$$

The max-$\boldsymbol{t}$ composition on the other hand realizes a so-called OR neuron which behaves dually to the AND neuron.

When put together the corresponding expressions give rise to some structure with a clear logical characteristics. The fuzzy data accomodated here modify the connections (viz. R and G) and create links within the structure. The structure of $\mathcal{R}$ is essentially characterized by an input layer $\mathcal{X} = \{x_1, \ldots, x_n\}$, an output layer $\mathcal{Y} = \{y_1, \ldots, y_n\}$, and the intermediate, hidden layer $\mathcal{Z}$, each one of them a finite set of nodes.

3 Solvability of coupled fuzzy relational equations

The basic connection between both the types of relation equations mentioned in (17) is given by the facts that

$$B = A \circ R \qquad \text{iff} \qquad \overline{B} = \overline{A} \# \overline{R}, \tag{20}$$

$$B = A \# R \qquad \text{iff} \qquad \overline{B} = \overline{A} \circ \overline{R}. \tag{21}$$

Here $\overline{A}$ denotes the complement of the fuzzy set A.

From the two types of structural possibilities mentioned in (18) and (19) we are lead to two types of cascaded structures described by different types of systems of relational equations. For these structures the input/output behaviour realizing the data pairs (16) is to be modelled either by a coupled system of fuzzy relation equations of the form

$$B_k = Z_k \# G \quad \text{and} \quad Z_k = A_k \circ R, \qquad k = 1, \ldots, N. \tag{22}$$

for a structure of the type (18), or for a structure of the type (19) by a coupled system of fuzzy relation equations of the form

$$B_k = Z_k \circ G \quad \text{and} \quad Z_k = A_k \# R, \qquad k = 1, \ldots, N. \tag{23}$$

Furthermore we restrict for simplicity to the case $t = \min$, but the extension to the general case is not an essential problem.

Using the binary operations α and its dual ε in $[0, 1]$ as well as their variants Ⓐ and Ⓔ for fuzzy sets (which produce fuzzy relations as results), cf. e.g. [Pedrycz 1989], one is able to translate the well-known solvability criterion of [Sanchez 1984] into the result:

Proposition 1 *A relational equation $B = A \# R$ for an unknown fuzzy relation R has a solution iff $\check{R} = A$ Ⓔ B is a solution of it; and if $\check{R} = A$ Ⓔ B is a solution it is the smallest one.*

With those results in mind one is able to discuss the case $N = 1$ of one pair of fuzzy relational equations (22) and (23) which means to consider one of the coupled pairs of fuzzy relational equations

$$B = Z \# G \quad \text{and} \quad Z = A \circ R, \tag{24}$$

$$B = Z \circ G \quad \text{and} \quad Z = A \# R. \tag{25}$$

Proposition 2 *A coupled pair of relational equations* (24) *has as solution the pair of fuzzy relations*

$$(G, R) = (Z \circledcirc B, A \circledast Z) \tag{26}$$

if and only if

$$\inf_z Z(z) \leq \inf_y B(y) \quad \textit{and} \quad \sup_z Z(z) \leq \sup_x A(x). \tag{27}$$

Here the "parameter" Z in the case that the input value A is a normal fuzzy set, i.e. has $\text{hgt}\,(A) = 1$, can easily be chosen in such a way that $Z(z_0) = 0$ for some $z_0 \in \mathcal{Z}$, which means that $\text{hgt}\,(\overline{Z}) = 1$. Even in the case $\text{hgt}\,(A) < 1$ it is not a problem to find some Z meeting conditions (27).

Hence, by (26) a whole family of solutions of the equations (24) is given.

For equations (25) we have the dual situation.

Our next problem of course is to extend these considerations to systems of coupled pairs of relational equations.

Actually it seems to be quite difficult to discuss solutions of system (23). There is a tour de force approach which uses even the pairwise disjointness of the intermediate fuzzy sets Z_i as an additional assumption. But this assumption does not seem to be reasonable at all.

For the other system (22) of coupled fuzzy relational equations the situation is much better because now the partial system

$$Z_k = A_k \circ R, \qquad k = 1, \ldots, N, \tag{28}$$

immediately refers to the input data of the process. For them it is much more reasonable to assume their pairwise disjointness

$$A_i \cap A_j = \emptyset \qquad \text{for all } 1 \leq i < j \leq N \tag{29}$$

than to assume that condition for the intermediate fuzzy sets. Therefore we decide to discuss the solvability of the system (22) under the additional assumption (29).

Then parallel we consider the systems of equations

$$\begin{aligned} Z_k &= A_k \circ R, \qquad k = 1, \ldots, N, & (30) \\ \overline{B_k} &= \overline{Z_k} \circ \overline{G}, \qquad k = 1, \ldots, N & (31) \end{aligned}$$

such that because of (21) both systems (30) and (31) together are equivalent with the complete system (22).

The previous remarks confirm us that the system (30) has well known solutions in case one supposes the disjointness (29) of the input data.

Fortunately, for the system (31) we can follow the same strategy using the fact that our intermediate fuzzy sets Z_k can be choosen in such a way that the fuzzy sets $\overline{Z_k}$, $k = 1, \ldots, N$ are pairwise disjoint, i.e.

$$Z_i \cup Z_j = \text{universal set over } \mathcal{Z} \qquad \text{for all } 1 \leq i < j \leq N. \qquad (32)$$

Therefore (29) and (32) give as a result the following

Proposition 3 *Assuming the pairwise disjointness* (29) *of the input data and additionally that the intermediate fuzzy sets Z_k, $k = 1, \ldots, N$ met the condition (32) then the whole system* (22) *has the "*MAMDANI*-type" solution*

$$(R_M, G_M) = \Big(\bigcup_{k=1}^{N} (A_k \times Z_k), \bigcap_{k=1}^{N} \overline{(\overline{Z_k} \times \overline{B_k})} \Big)$$

as well as the solution

$$(\hat{R}, \check{G}) = \Big(\bigcap_{k=1}^{N} (A_k @ Z_k), \bigcup_{k=1}^{N} (Z_k \circledcirc B_k) \Big).$$

All those results, however, have to refer to additional assumptions – assumptions which in some cases seem to be very restrictive, and a little less restrictive in other cases. Anyway, those assumptions usually may be hard to be satisfied in real applications. In this respect, the situation with our coupled systems of fuzzy relational equations is not simpler as for the usual types of fuzzy relational equations as discussed e.g. in [diNola et al. 1989], [Gottwald 1986], [Gottwald 1993].

References

[diNola et al. 1988] A.diNola, W.Pedrycz, S.Sessa: Fuzzy relation equations with equality and difference composition operators. *Fuzzy Sets Syst.* **25** (1988), 205 - 215.

[diNola et al. 1989] A.diNola, S.Sessa, W.Pedrycz, E.Sanchez: *Fuzzy Relation Equations and Their Applications to Knowledge Engineering.* Theory and Decision Library, ser. D, Kluwer: Dordrecht, 1989.

[Gottwald 1984] S.Gottwald: On the existence of solutions of systems of fuzzy equations. *Fuzzy Sets Syst.* **12** (1984), 301 - 302.

[Gottwald 1986] S.Gottwald: Characterizations of the solvability of fuzzy equations. *Elektron. Informationsverarb. Kybernetik* EIK **22** (1986), 67 - 91.

[Gottwald 1993] S.Gottwald: *Fuzzy Sets and Fuzzy Logic.* Vieweg: Braunschweig and Teknea: Toulouse, 1993.

[Ohsato, Sekiguchi 1983] A.Ohsato, T.Sekiguchi: Convexly combined form of fuzzy relational equations and its application to knowledge representation. In *Proc. Intern. Conf. Systems, Man and Cybernet., Bombay 1983/84*; vol. 1, Bombay - New Delhi (IEEE India Council), 294 - 299.

[Ohsato, Sekiguchi 1985] A.Ohsato, T.Sekiguchi: Maximin solution of the convexly combined form of composite fuzzy relation equations [in Japanese]. *Transact. Soc. Instrument and Control Engineers* (Japan) **21** (1985), 423 - 428.

[Pedrycz 1985] W.Pedrycz: Applications of fuzzy relational equations for methods of reasoning in presence of fuzzy data. *Fuzzy Sets Syst.* **16** (1985), 163 - 175.

[Pedrycz 1989] W.Pedrycz: *Fuzzy Control and Fuzzy Systems.* Research Stud. Press: Taunton and Wiley: New York, 1989.

[Sanchez 1984] E.Sanchez: Solution of fuzzy equations with extended operations. *Fuzzy Sets Syst.* **12** (1984), 237 - 248.

[Zadeh 1973] L.A.Zadeh: Outline of a new approachto the analysis of complex systems and decision processes. *IEEE Trans. Systems, Man and Cybernet.* **SMC-3** (1973), 28 - 44.

5.3. Monoidal Logic

Ulrich Höhle

Abstract

Monoidal logic is the a common framework for intuitionistic logic, Lukasiewicz logic and to a ceratin extent for Girard's commutative logic. Soundness and completeness of the corresponding predicate calculi are verified.

Keywords. Predicate calculus w.r.t. non-classical logics, Lindenbaum algebra, Heyting algebras, MV–algebras, Girard–monoids, t-norms.

Introduction

The aim of this paper is to present the *syntactic basis* of fuzzy logic. For this purpose we specify the monoidal predicate calculus which turns out to be sound and complete. As special cases we obtain the intuitionistic predicate calculus, Girard's integral, commutative predicate calculus and Lukasiewicz predicate calculus.

1 Monoidal predicate calculus

Let $\mathcal{L}$ be a formalized language of first order and $\{\neg, \wedge, \vee, \rightarrow, \otimes\}$ be the set of logical symbols where $\neg$ is a unary and the remaining symbols are binary operations. The *logical axioms* of *monoidal logic* are the following axiom schemes

(T_1) $((\alpha \rightarrow \beta) \rightarrow ((\beta \rightarrow \gamma) \rightarrow (\alpha \rightarrow \gamma)))$ (Syllogism Law)

(T_2) $(\alpha \rightarrow (\alpha \vee \beta))$

(T_3) $(\beta \rightarrow (\alpha \vee \beta))$

(T_4) $((\alpha \rightarrow \gamma) \rightarrow ((\beta \rightarrow \gamma) \rightarrow ((\alpha \vee \beta) \rightarrow \gamma)))$

(T_5) $((\alpha \wedge \beta) \rightarrow \alpha)$

(T_5') $((\alpha \otimes \beta) \rightarrow \alpha)$

(T_6) $((\alpha \wedge \beta) \rightarrow \beta)$

(T_6') $((\alpha \otimes \beta) \rightarrow (\beta \otimes \alpha))$

(T_6'') $((\alpha \otimes (\beta \otimes \gamma)) \rightarrow ((\alpha \otimes \beta) \otimes \gamma))$

(T_7) $((\gamma \rightarrow \alpha) \rightarrow ((\gamma \rightarrow \beta) \rightarrow (\gamma \rightarrow (\alpha \wedge \beta))))$

(T_8) $((\alpha \rightarrow (\beta \rightarrow \gamma)) \rightarrow ((\alpha \otimes \beta) \rightarrow \gamma))$ (Importation Law)

(T_9) $(((\alpha \otimes \beta) \rightarrow \gamma) \rightarrow (\alpha \rightarrow (\beta \rightarrow \gamma)))$ (Exportation Law)

(T_{10}) $((\alpha \otimes \neg\alpha) \rightarrow \beta)$ (Duns Scotus)

(T_{11}) $((\alpha \rightarrow (\alpha \otimes \neg\alpha)) \rightarrow \neg\alpha)$

Further we assume the usual quantifier axioms – i.e. for every well formed formula α and for every term τ for which the variable v is free in α the following expressions are axiom schemes

(UI) $((\forall v)\alpha \rightarrow \alpha(v/\tau))$

(EG) $(\alpha(v/\tau) \rightarrow (\exists v)\alpha)$

The *monoidal,predicate calculus* PC is the usual predicate calculus based on the logical axioms (T_1) – (T_5) , (T_5'), (T_6), (T_6'), (T_6''), (T_7) – (T_{11}) , on the quantifier axioms (UI) , (EG) , on *Modus Ponens* (MP) and the usual quantifier rules

($\forall$) From $(\alpha \rightarrow \beta)$ infer $(\alpha \rightarrow (\forall v)\beta)$ provided v is not free in α .

($\exists$) From $(\alpha \rightarrow \beta)$ infer $((\exists v)\alpha \rightarrow \beta)$ provided v is not free in β .

We apply the usual notations – e.g. if α is provable, then this situation is denoted by $\vdash \alpha$.

Lemma 1.1 *Let PC be the monoidal predicate calculus. Then for all* $\alpha, \beta, \gamma \in \mathcal{L}$ *the following relations hold*

(i) $\vdash \quad (\alpha \rightarrow \alpha)$

(ii) $\vdash \quad ((\alpha \otimes (\alpha \rightarrow \beta)) \rightarrow \beta)$

(iii) $\vdash \quad (\alpha \rightarrow (\beta \rightarrow (\beta \otimes \alpha)))$

(iv) If $\vdash \alpha$, then $\vdash \quad (\beta \rightarrow (\beta \otimes \alpha))$

(v) $\vdash \quad (\alpha \rightarrow (\beta \rightarrow \alpha))$

(vi) $\vdash \quad ((\beta \rightarrow \gamma) \rightarrow ((\alpha \otimes \beta) \rightarrow (\alpha \otimes \gamma)))$

Proof. Because of the exportation law (T_9) the formula

$$(((((\alpha \otimes \alpha) \rightarrow \alpha) \otimes \alpha) \rightarrow \alpha) \rightarrow (((\alpha \otimes \alpha) \rightarrow \alpha) \rightarrow (\alpha \rightarrow \alpha)))$$

is provable. Then (i) follows from $(T_1), (T_5'), (T_6')$ and several applications of (MP) . The relation (iv) is a consequence of (iii) . Further (ii), (iii) and (v) follow immediately from (i), $(T_1); (T_5'), (T_6'), (T_8)$ and (T_9) . In oder to verify (vi) we first infer from (ii), (iii), (T_1) and (T_6') that the formula

$$((\beta \otimes (\beta \rightarrow \gamma)) \rightarrow (\alpha \rightarrow (\alpha \otimes \gamma)))$$

is provable. Hence (vi) follows from $(T_1), (T_6'), (T_6''), (T_8)$ and (T_9) .
□

An important consequence of the syllogism law and the assertion (i) is the fact that the relation $\triangleright$ defined by

$$\alpha \triangleright \beta \quad \Longleftrightarrow \quad \vdash \quad (\alpha \rightarrow \beta)$$

is a preorder on the set $\mathcal{L}$ of all well formed formulas. If $\sim$ is the equivalence relation associated with $\triangleright$ (i.e. $\alpha \sim \beta \iff \vdash (\alpha \rightarrow \beta)$ and $\vdash (\beta \rightarrow \alpha)$), then we can consider the quotient $L = \mathcal{L}/\sim$ of all equivalence classes of logically equivalent formulas. In particular $\triangleright$ induces a partial ordering $\leq$ on L . Referring to $(T_2) - (T_7), (T_{10})$ and (v) it is easy to see that $(L, \leq)$ is a lattice with the top elemnt $1 = \{\alpha \in L \mid \vdash \alpha\}$ .Because of $(T_1), (T_6')$ and (vi) the logical symbol $\otimes$ defines a binary operation $*$ on L as follows

$$[\alpha] * [\beta] \quad = \quad [\alpha' \otimes \beta'] \qquad \text{where} \quad \alpha' \in [\alpha], \quad \beta' \in [\beta] \quad .$$

¿From the axioms $(T_5'), (T_6'), (T_6''), (T_8), (T_9)$ and (iv) we conclude that $(L, \leq, *)$ is an integral, residuated, commutative l–monoid ([2]); i.e. $(L, \leq)$ is a lattice with top and bottom element, $(L, *)$ is a commutative monoid in which the unity coincides with the top element 1, and there exists a further binary operation $\rightarrow$ satisfying the following axiom

(AD) $\quad \alpha * \beta \leq \gamma \Longleftrightarrow \alpha \leq \beta \rightarrow \gamma$.

In accordance with the terminology introduced by H. Rasiowa and R. Sikorski ([12]) the triple $(L, \leq, *)$ is also called the *Lindenbaum algebra* of the monoidal predicate calculus. Moreover we can show in the usual way that the following relations hold

$$[(\forall v)\alpha] = \bigwedge_{\tau}[\alpha(v/\tau)] \quad , \quad [(\exists v)\alpha] = \bigwedge_{\tau}[\alpha(v/\tau)]$$

Before we continue the discussion of the axioms of monoidal logic we first make a short digression into the theory of integral, commutative, residuated l–monoids.

2 Remarks on integral, commutative, residuated l–monoids

An integral, commutative, residuated l–monoid $(L, \leq, *)$ is said to be *divisible* iff for every pair $(\alpha, \beta) \in L$ with $\alpha \leq \beta$ there exists $\gamma \in L$ s.t. $\alpha = \beta * \gamma$. An integral, commutative, residuated l–monoid is called a *Girard–monoid* iff the "negation" is an involution - i.e. $(\alpha \rightarrow 0) \rightarrow 0 = \alpha$. An MV–algebra is an integral, divisible, commutative Girard–monoid.

Examples 2.1 (Real unit interval) Any left–continuous t-norm (cf. [14]) determines on the real unit interval [0,1] the strucuture of an integral, commutative, residuated l–monoid. Further the *nilpotent minimum* T_0 (cf. [4])

$$T_0(x, y) = \begin{cases} Min(x, y) & : \; 1 < x + y \\ 0 & : \; x + y \leq 1 \end{cases}$$

defines on [0,1] the structure of an integral, commutative Girard–monoid which is not an MV–algebra.
Integral, divisible, commutative, residuated l–monoidal structures on [0,1] are exactly given by continuous t–norms. In particular Lukasiewicz arithmetic conjunction T_m (cf. [5])

$$T_m(x, y) = Max(x + y - 1, 0)$$

induces an MV–algebra structure on [0,1]. It is interesting to see that the "negations" w.r.t. T_0 and w.r.t. T_m coincide with L.A. Zadeh's negation n (cf. [17])

$$n(x) \quad = \quad 1 - x \quad .$$

Finally we remark that the usual product determines an integral, divisible, monoidal structure on [0,1] which is neither a Heyting algebra nor an MV–algebra.

Theorem 2.2 ([10]) *The* MacNeille *completion of an integral, commutative, residuated l–monoid is again an integral, commutative, residuated l–monoid. In particular the structure of Heyting algebras as well as of integral, commutative Girard–monoids are preserved under the MacNeille completion.*

It is an important observation that the MacNeille completion does *not* preserve the *divisibility* of integral, commutative, residuated l–monoids. In this respect we have the following

Theorem 2.3 ([10]) *Let $M = (L, \leq, *)$ be an MV–algebra. Then the following assertions are equivalent*

(i) The MacNeille completion of M is again an MV–algebra.

(ii) $\forall \alpha \in L$ with $\alpha \neq 1$ $\exists n \in I\!N$ s.t.[1] $(\alpha \to 0) \to \alpha^n \neq 1$.

It is well known (cf. [1], [10]) that the condition (ii) in the previous theorem 2.3 is equivalent to the semi–simplicity of MV–algebras.
An MV–algebra is called σ–complete iff the underlying lattice $(L, \leq)$ is σ–complete –i.e. countable joins and meets exist in $(L, \leq)$.

Proposition 2.4 (Sufficient Condition for Semi–simplicity ([1])) *Every σ–complete MV–algebra is semi-simple.*

Theorem 2.5 (Tarski's Lemma ([10])) *Let $M = (L, \leq, *)$ be a σ–complete MV–algebra and $\alpha_0 \in L$ with $\alpha_0 \neq 1$. Further let $\{A_n | n \in I\!N\}$ be a countable family of countable subsets A_n of L. Then there exists an MV–algebrahomo-
morphism $h : M \longmapsto ([0,1], \leq, T_m)$ provided with the following properties*

(1) $h(\alpha_0) \neq 1$,

(2) $\inf_{\beta \in A_n} h(\beta) \quad = \quad h(\wedge A_n)$ for all $n \in I\!N$

[1] α^n denotes the n–th power of α w.r.t. $*$.

3 Special Cases of Monoidal Logics

Remark 3.1 We add to the system of logical axioms of monoidal logic the *law of idempotency* – i.e. the axiom schema

(T_{12}') $\qquad (\alpha \rightarrow (\alpha \otimes \alpha))$.

Then the logical symbols $\otimes$ and $\wedge$ are logically equivalent and the Lindenbaum algebra is a *Heyting algebra* ([11]). In this context the extended axiom system reduces to the axioms of *intuitionistic logic* and PC coincides with the intuitionistic predicate calculus.

Remark 3.2 If we adjoin to the axioms of monoidal logic the *law of double negation* – i.e. the axiom scheme

(T_{12}'') $\qquad (\neg\neg\alpha \rightarrow \alpha)$,

then the Lindenbaum algebra is an *integral, commutative Girard–monoid*, and we arrive at Girard's integral, commutative *linear logic* ([6]).

Remark 3.3 If we add to the axiom system of monoidal logic the *law of divisibility* – i.e. the axiom schema

(T_{12}''') $\qquad ((\alpha \wedge \beta) \rightarrow (\alpha \otimes (\alpha \rightarrow \beta)))$

and the *law of double negation* (T_{12}'') , then the Lindenbaum algebra is a divisible Girard–monoid – i.e. an MV–*algebra* (cf. [3]). In this context the extended axiom system reduces to the *Wajsberg axioms* of *Lukasiewicz logic* ,namely to the following well known system

$(\alpha \rightarrow (\beta \rightarrow \alpha))$ (Affirmation of the Consequent)

$((\alpha \rightarrow \beta) \rightarrow ((\beta \rightarrow \gamma) \rightarrow (\alpha \rightarrow \gamma)))$ (Syllogism Law)

$(((\alpha \rightarrow \beta) \rightarrow \beta) \rightarrow ((\beta \rightarrow \alpha) \rightarrow \alpha))$

$((\neg\alpha \rightarrow \neg\beta) \rightarrow (\beta \rightarrow \alpha))$ (Contraposition Law)

In particular we obtain *Lukasiewicz predicate calculus* ([9]).

Finally, if we adjoin to the axioms of monoidal logic the axiom (T_{12}'), (T_{12}''),
(T_{12}), then we arrive at the *classical* predicate calculus; and the Lindenbaum algebra is a *Boolean algebra* ([16])

4 Soundness and completeness

Let M $= (L, \leq, *)$ be an integral, commutative cl–monoid – i.e. an integral, commutative, residuated l–monoid such that the underlying lattice $(L, \leq)$ is complete ([2]). A M–*valued intepretation* of PC consists of an univers X and two mappings Φ and Ψ where

- Φ assigns to every n–ary functional symbol f an n–ary operation $\Phi(f) : X^n \longmapsto X$ on X ,
- Ψ assigns to every n–ary predicate symbol p a map $\Psi(p) : X^n \longmapsto L$.

Further let $\mathcal{V}$ be the set of all individual variables v and $\mathcal{T}$ the set of all terms τ. On $\mathcal{T}$ we consider the algebra structure induced by the functional symbols $f_1, f_2, \ldots$. Then every *valuation* $\nu : \mathcal{V} \longmapsto X$ can uniquely be extended to an algebra–homomorphism $h_\nu : \mathcal{T} \longmapsto X$. In particular we have

$$h_\nu(f(\tau_1, \ldots, \tau_n)) \quad = \quad [\Phi(f)](h_\nu(\tau_1), \ldots, h_\nu(\tau_n)) \quad .$$

Moreover every valuation ν determines a map $\Theta_\nu : \mathcal{L} \longmapsto L$ as follows

1. $\Theta_\nu(p(\tau_1, \ldots, \tau_n)) \quad = \quad [\Psi(p)](h_\nu(\tau_1), \ldots, h_\nu(\tau_n))$ (atomic formulas)

2. $\Theta_\nu(\neg\alpha) \quad = \quad (\Theta_\nu(\alpha) \to 0) \quad ,$

3. $$\begin{aligned} \Theta_\nu(\alpha \to \beta) &= \Theta_\nu(\alpha) \to \Theta_\nu(\beta) \\ \Theta_\nu(\alpha \otimes \beta) &= \Theta_\nu(\alpha) * \Theta_\nu(\beta) \quad , \end{aligned}$$

4. $$\begin{aligned} \Theta_\nu(\alpha \wedge \beta) &= \Theta_\nu(\alpha) \wedge \Theta_\nu(\beta) \\ \Theta_\nu(\alpha \vee \beta) &= \Theta_\nu(\alpha) \vee \Theta_\nu(\beta) \quad , \end{aligned}$$

5. $$\begin{aligned} \Theta_\nu((\forall v)\alpha) &= \bigwedge_{x \in X} \Theta_{\nu_x}(\alpha) \\ \Theta_\nu((\exists v)\alpha) &= \bigvee_{x \in X} \Theta_{\nu_x}(\alpha) \quad , \end{aligned}$$

$$\text{where} \quad \nu_x(v) = \begin{cases} \nu(v): & v \neq x \\ x: & v = x \end{cases}$$

We use the following terminology : Let M be an integral, commutative cl–monoid; a well-formed formula $\alpha \in \mathcal{L}$ is said to be M–*valid* iff

$\Theta_\nu(\alpha) = 1$ for all M–valued intepretations of PC and for all valuations ν. $\alpha \in \mathcal{L}$ is called **ICL**–*valid* iff α is M–valid for all integral, commutative cl–monoids M. $\alpha \in \mathcal{L}$ is **HA**–*valid* iff α is Ω–valid for all complete Heyting algebras Ω ([7]). $\alpha \in \mathcal{L}$ is **CGQ**–*valid* iff α is M–valid for all integral, commutative, complete Girard–monoids (i.e. for all integral, commutative Girard quantales ([13]) M . $\alpha \in \mathcal{L}$ is **MV**–*valid* iff α is M–valid for all complete MV–algebras M.

¿From Theorem 2.2 we obtain immediately :

Theorem 4.1 (Soundness and Completeness) *The monoidal predicate calculus is sound and complete – i.e. α is provable iff α is* **ICL**–*valid. The predicate calculus w.r.t. Girard's integral commutative linear logic is sound and complete – i.e. α is provable w.r.t. Girard's integral, commutative, linear logic (cf. Remark 3.2) iff α is* **CGQ**–*valid. The intuitionistic predicate calculus is sound and complete –i.e. α is provable w.r.t. the intuitionistic logic (cf. Remark 3.1) iff α is* **HA**–*valid.*

Referring to Scarpellini's result (cf. [15]) Łukasiewicz predicate calculus is sound but incomplete. In order to overcome this difficulty we add an additional, infinitary inference rule

(R) From $((\alpha \to 0) \to \alpha^n)$ for all natural numbers $n \in \mathbb{N}$, infer α

to Łukasiewicz predicate calculus. This approach leads to a *modified* version of Łukasiewicz predicate calculus denoted by ŁPC* . Since the Lindenbaum algebra of ŁPC* is semi–simple, the MacNeille completion preserves its MV-algebra structure. Applying the version of Tarski's Lemma stated in Theorem 2.5 we can follow the strategy of H. Rasiowa and R. Sikorski's proof of Gődel's completeness theorem and obtain the important

Theorem 4.2 *Let α be a well-formed formula. Then the following assertions are equivalent*

(i) α is provable in ŁPC* .

(ii) α is MV–valid.

(iii) α is $([0,1], \leq, T_m)$–valid.

In view of Theorem 4.2 LPC* is sound and complete. In particular a well-formed formula α is provable in LPC* if and only if for every [0,1]-valued interpretation and for every valuation ν the value of Θ_ν at α attains 1 (i.e $\Theta_\nu(\alpha) = 1$).

Therefore we view the calculus LPC* as the *syntactic counterpart* of *fuzzy logic* .

Addition. The equivalence between the assertions (i) and (iii) in Theorem 4.2 was first established by L. Hay [9]

5 Concluding remark

We finish this paper with a discussion of Axiom (T_5') . It is not difficult to see that (T_5') forces the integrality of the monoidal predicate calculus – i.e. that unity and top element of the corresponding Lindenbaum algebra coincide. If the reader does not like this limitation, we emphasize that he can easily over comme this obstacle – e.g. we can introduce a *logical constant* 1 and replace (T_5'),(T_{10}) and (T_{11}) by the following axiom schemes

$(\hat{T}_5)$ $(\alpha \to (1 \otimes \alpha))$

$(\hat{\hat{T}}_5)$ $((1 \otimes \alpha) \to \alpha)$

$(\hat{T}_{10})$ $((\alpha \otimes \neg\alpha) \to \neg 1)$

$(\hat{T}_{11})$ $((\alpha \to \neg 1) \to \neg\alpha)$.

In this situation the Lindenbaum algebra is a commutative, residuated l-monoid in which the unity [1] is different from the top element. The subsequent relation is valid

$$\vdash \alpha \quad \text{iff} \quad [1] \le [\alpha] \quad \text{iff} \quad \vdash (1 \to \alpha) \quad .$$

Now the semantic side requires general, commutative cl-monoids M (cf. [2]), and , if 1 is the unity of M, then M-validity of a well-formed formula α means of course that the inequality $1 \le \Theta_\nu(\alpha)$ holds for all M-valued interpretions of PC and for all valuations ν.
Finally, if we add the law of double negation (cf. Remark 3.2), then we arrive at Girard's (*non-integral*) commutative,linear logic without modalities ([6]).

References

[1] L.P. Belluce . *Semi-simple and complete MV-algebras*, Algebra Universalis **29** (1992), 1 - 9 .

[2] G. Birkhoff . *Lattice Theory* , Amer. Math. Soc. Colloquium Publications, third edition (Amer. Math. Soc., RI, 1973).

[3] C.C. Chang . *Algebraic analysis of many valued logics*, Trans. Amer. Math. Soc. **88** (1958), 467 - 490 .

[4] J.C. Fodor . *Contrapositive symmetry of fuzzy implications*, Technical Report 1993/1, Eőtvős Loránd University, Budapest 1993.

[5] O. Frink . *New algebras of logic*, Amer. Math. Monthly **45** (1938), 210 - 219

[6] J.Y. Girard . *Linear logic*, Theor. Comp. Sci. **50** (1987), 1-102.

[7] R. Goldblatt . *Topoi : The Categorial Analysis of Logic* (North-Holland , Amsterdam 1979).

[8] S. Gottwald . *Mehrwertige Logik* (Akademie-Verlag, Berlin 1989).

[9] L.S. Hay . *Axiomatization of infinite-valued predicate calculus*, Journal of Symbolic Logic **28** (1963), 77-86 .

[10] U. Hőhle . *Commutative, residuated l-monoids* in "Non-Classical Logics and Their Applications to Fuzzy Subsets: A Handbook of the Mathematical Foundations of Fuzzy Set Theory" (Eds. U. Hőhle and E.P. Klement) (Kluwer 1995).

[11] P.T. Johnstone . *Stone Spaces* (Cambridge University Press, Cambridge 1982).

[12] H. Rasiowa and R. Sirkoski . *A proof of the completeness theorem of Gődel*, Fundamenta Math. **37** (1950), 193-200.

[13] K.I. Rosenthal . *Quantales and Their Applications* , Pitman Research Notes in mathematics **234** (Longman, Burnt Mill, Harlow 1990).

[14] B. Schweizer and A. Sklar . *Probabilistic Metric Spaces* (North-Holland , Amsterdam 1983).

[15] B. Scarpellini . *Die Nichtaxiomatisierbarkeit des unendlichen Prădikaten-kalkűls von Łukasiewicz*, Journal of Symolic Logic **27** (1962), 159–170.

[16] R. Sikorski . *Boolean algebras* (Springer–Verlag, Berlin 1964).

[17] L.A. Zadeh . *Fuzzy sets*, Information and Control **8** (1965), 338 – 353 .

5.4. Interpolation and Approximation of Real Input–Output Functions Using Fuzzy Rule Bases

Peter Bauer, Erich Peter Klement,
Albert Leikermoser and Bernhard Moser

Abstract

It is shown how fuzzy controllers, in particular the Mamdani and Sugeno controller, can be used to interpolate and approximate control functions, i.e., input–output functions which assign to each input value a real output value.

1 Fuzzy controllers

In this paper we shall restrict ourselves to the two most important and most widely used fuzzy controllers, the Mamdani [7] and the Sugeno controller [8]. For a general overview on fuzzy control, see, e.g., [3].

We start with the Mamdani controller which uses fuzzy sets both for input and output and, therefore, needs a defuzzification in order to produce an input–output function.

Definition 1 Let X be an arbitrary input space, let $A_1, A_2, \ldots, A_n$ and $B_1, B_2, \ldots, B_n$ be normalized fuzzy subsets with Borel–measurable membership functions of X and $\mathbf{R}^m$, respectively, let T be a Borel–measurable t-norm, and consider the rulebase $(i = 1, 2, \ldots, n)$

$$\text{IF} \quad x \text{ is } A_i \quad \text{THEN} \quad \mathbf{u} \text{ is } B_i.$$

Then, provided we have $\int \mu_R(x, \mathbf{u})\, d\mathbf{u} > 0$, the *Mamdani controller* defines the following input–output function $F_M : X \to \mathbf{R}^m$

$$F_M(x) = \frac{\displaystyle\int_{\mathbf{R}^m} \mu_R(x, \mathbf{u}) \cdot \mathbf{u}\, d\mathbf{u}}{\displaystyle\int_{\mathbf{R}^m} \mu_R(x, \mathbf{u})\, d\mathbf{u}}, \tag{1}$$

where the membership function μ_R of the fuzzy relation R on $X \times \mathbf{R}^m$ is given by

$$\begin{aligned} \mu_R(x, \mathbf{u}) \; = \; & \max\big[T\big(\mu_{A_1}(x), \mu_{B_1}(\mathbf{u})\big), T\big(\mu_{A_2}(x), \mu_{B_2}(\mathbf{u})\big), \ldots, \\ & T\big(\mu_{A_n}(x), \mu_{B_n}(\mathbf{u})\big)\big]. \end{aligned} \tag{2}$$

In a strict mathematical sense, the measurability requirements are necessary for (1) to be well-defined; in practical situations, these hypotheses, however, are usually satisfied.

In Definition 1 we have implicitly chosen a special defuzzification method, namely, the so-called *center of area*, which is basically contained in equation (1). We only mention that there are also other methods of defuzzification, e.g., the *mean of maximum*.

In most practical examples, the t-norm used for the Mamdani controller is either the *minimum* $T_{\mathbf{M}}$ or the *product* $T_{\mathbf{P}}$; in the first case this is also referred to as *max-min-inference*, in the latter case as *max-prod-inference* or *max-dot-inference*.

The second important type of fuzzy controllers is the so-called Sugeno controller which uses crisp values in the output space. In a way this means that the inference has a built-in defuzzification.

Definition 2 Let X be an input space, let $A_1, A_2, \ldots, A_n$ be normalized fuzzy subsets of X with $\sum \mu_{A_i}(x) > 0$ for all $x \in X$, and $f_1, f_2, \ldots, f_n$ be functions from X to $\mathbf{R}^m$, and consider the rulebase $(i = 1, 2, \ldots, n)$

$$\text{IF} \quad x \text{ is } A_i \quad \text{THEN} \quad \mathbf{u} = f_i(x).$$

Then the *Sugeno controller* defines the following input-output function $F_S : X \to \mathbf{R}^m$

$$F_S(x) = \frac{\sum \mu_{A_i}(x) \cdot f_i(x)}{\sum \mu_{A_i}(x)}. \tag{3}$$

In the special situation, when for $i = 1, 2, \ldots, n$ the functions f_i are constant, i.e., $f_i(x) = \mathbf{u}_i$, the Sugeno controller can be considered as a special case of the Mamdani controller:

Theorem 1 *Let X be an input space, let T be a t-norm, let $A_1, A_2, \ldots, A_n$ be normalized fuzzy subsets of X with $\sum \mu_{A_i}(x) > 0$ for all $x \in X$, let $\mathbf{u}_1, \mathbf{u}_2, \ldots, \mathbf{u}_n$ be different elements of $\mathbf{R}^m$, consider the rulebase $(i = 1, 2, \ldots, n)$*

$$\text{IF} \quad x \textit{ is } A_i \quad \text{THEN} \quad \mathbf{u} = \mathbf{u}_i,$$

and let F_S be the input-output function of the corresponding Sugeno controller. Then there exists a Mamdani controller with input space X such that its corresponding input-output function F_M coincides with F_S.

PROOF. For each positive ε and each $i = 1, 2, \ldots, n$ consider the closed ε-ball $B_i^{(\varepsilon)}$ with center $\mathbf{u}_i = \left(u_i^{(1)}, u_i^{(2)}, \ldots, u_i^{(m)}\right) \in \mathbf{R}^m$

$$\begin{aligned} B_i^{(\varepsilon)} = \{ &\left(u^{(1)}, u^{(2)}, \ldots, u^{(m)}\right) \in \mathbf{R}^m \mid \\ &\max\left\{\left|u^{(1)} - u_i^{(1)}\right|, \left|u^{(2)} - u_i^{(2)}\right|, \ldots, \left|u^{(m)} - u_i^{(m)}\right|\right\} \leq \varepsilon\}, \end{aligned} \quad (4)$$

and the rulebase $(i = 1, 2, \ldots, n)$

$$\text{IF} \quad x \text{ is } A_i \quad \text{THEN} \quad \mathbf{u} \text{ is } B_i^{(\varepsilon)}.$$

There exists an $\varepsilon_0 > 0$ such that for all $i \neq j$ we have $B_i^{(\varepsilon_0)} \cap B_j^{(\varepsilon_0)} = \emptyset$, and the fuzzy relation $R^{(\varepsilon_0)}$ defined in (2) induced by the corresponding Mamdani controller becomes

$$\begin{aligned} \mu_{R^{(\varepsilon_0)}}(x, \mathbf{u}) &= \max\left\{T\left(\mu_{A_i}(x), \mu_{B_i^{(\varepsilon_0)}}(\mathbf{u})\right) \mid i \in \{1, 2, \ldots, n\}\right\} \\ &= \begin{cases} \mu_{A_i}(x) & \text{if } x \in B_i^{(\varepsilon_0)}, \\ 0 & \text{otherwise.} \end{cases} \end{aligned}$$

The resulting input–output function $F_M^{(\varepsilon_0)}$ then yields

$$\begin{aligned} F_M^{(\varepsilon_0)}(x) &= \frac{\int_{\mathbf{R}^m} \mu_{R^{(\varepsilon_0)}}(x, \mathbf{u}) \cdot \mathbf{u}\, d\mathbf{u}}{\int_{\mathbf{R}^m} \mu_{R^{(\varepsilon_0)}}(x, \mathbf{u})\, d\mathbf{u}} \\ &= \frac{\sum \left[\int_{B_i^{(\varepsilon_0)}} \mu_{A_i}(x) \cdot \mathbf{u}\, d\mathbf{u}\right]}{(2\varepsilon_0)^m \cdot \sum \mu_{A_i}(x)} \\ &= \frac{\sum \left[\mu_{A_i}(x) \cdot \int_{B_i^{(\varepsilon_0)}} \mathbf{u}\, d\mathbf{u}\right]}{(2\varepsilon_0)^m \cdot \sum \mu_{A_i}(x)} \\ &= \frac{(2\varepsilon_0)^m \cdot \sum \mu_{A_i}(x) \cdot \mathbf{u}_i}{(2\varepsilon_0)^m \cdot \sum \mu_{A_i}(x)}, \\ &= F_S(x). \end{aligned}$$

showing that $F_M^{(\varepsilon_0)} = F_S$. □

In $\mathbf{R}^m$ we can consider crisp points $\mathbf{x}_0$, i.e., one–point sets $\{\mathbf{x}_0\}$, as limits of closed ε-balls $B_i^{(\varepsilon)}$ as ε goes to zero. Therefore Theorem 1 also says that Sugeno controllers (with constant functions f_i) are limits of suitable Mamdani controllers, a result which holds for other defuzzification methods too.

2 Uniform approximation

It is rather natural to ask which input–output functions can be modeled or approximated using fuzzy controllers. There are several results showing that Sugeno controllers can be used to approximate in a uniform way continuous input–output functions defined on a compact subset of $\mathbf{R}^m$. The following theorem can be found in [9], where also the complete proof is presented:

Theorem 2 *Let $D = [a^{(1)}, b^{(1)}] \times [a^{(2)}, b^{(2)}] \times \dots [a^{(m)}, b^{(m)}]$ be a compact subset of $\mathbf{R}^m$, $f : D \to \mathbf{R}$ a continuous function and $\varepsilon > 0$. Then there exist real numbers $u_1, u_2, \dots, u_K$, fuzzy subsets $A_k^{(i)}$ of $[a^{(i)}, b^{(i)}]$ $(i = 1, 2, \dots, m;\ k = 1, 2, \dots, K)$ with Gaussian membership functions*

$$\mu_{A_k^{(i)}}(x) = \alpha_k^{(i)} \cdot e^{-\frac{1}{2}\left(\frac{x - x_k^{(i)}}{\sigma_k^{(i)}}\right)^2}, \tag{5}$$

with $\alpha_k^{(i)} \in (0, 1]$, $x_k^{(i)} \in \mathbf{R}$, $\sigma_k^{(i)} \in (0, \infty)$, and a rule base $(k = 0, 1, \dots, K)$

$$\text{IF } x^{(1)} \textit{ is } A_k^{(1)} \text{ AND } x^{(2)} \textit{ is } A_k^{(2)} \text{ AND } \dots \text{ AND } x^{(m)} \textit{ is } A_k^{(m)} \quad \text{THEN} \quad u = u_k,$$

such that the input–output function F_S of the corresponding Sugeno controller (using the t-norm $T_\mathbf{P}$ for the Cartesian product) satisfies

$$\sup\left\{|F_S(x) - f(x)| \;\middle|\; x \in D\right\} < \varepsilon. \tag{6}$$

¿From the point of view of a designer of a controller, Theorem 2 does not help a lot: its proof relies on the classical Stone–Weierstraß Theorem and is purely existential in nature, as it provides no hint at all which Sugeno controller should be chosen to approximate a concrete input–output function. Also, if the tolerances ε go to zero, the number of rules in general will not be bounded. Theorem 2 only states that, given a certain tolerance ε, there exists always a Sugeno controller whose input–output function F_S deviates from the given function less than the tolerance ε.

3 Interpolation by fuzzy systems

Since the result of Theorem 2 is only of limited practical relevance, it is important to know which input–output functions can be modeled by fuzzy systems. In the case of a one–dimensional input it is possible to show that a large class of functions can be realized by Sugeno controllers [6, 1].

Theorem 3 *Let $f : [\underline{a}, \overline{a}] \to \mathbf{R}$ be a continuous, piecewise monotone function. Then there exist normalized fuzzy subsets $A_0, A_1, \ldots, A_n$ of $[\underline{a}, \overline{a}]$, numbers $u_0, u_1, \ldots, u_n \in \mathbf{R}$, and a rulebase $(i = 0, 1, \ldots, n)$*

$$\text{IF} \quad x \text{ is } A_i \quad \text{THEN} \quad u = u_i$$

such that the input-output function F_S of the corresponding Sugeno controller coincides with f.

PROOF. Choose a partition of $[\underline{a}, \overline{a}]$,

$$\underline{a} = a_0 < a_1 < \cdots < a_{n-1} < a_n = \overline{a},$$

such that each restriction $f/_{[a_{i-1}, a_i]}$ $(i = 1, 2, \ldots, n)$ is a monotone function, and define the functions $f_i : [a_{i-1}, a_i] \to \mathbf{R}$ by

$$f_i(x) = \begin{cases} \dfrac{f(x) - f(a_i)}{f(a_{i-1}) - f(a_i)} & \text{if } f(a_{i-1}) \neq f(a_i), \\ 1 & \text{if } f(a_{i-1}) = f(a_i). \end{cases}$$

Next we introduce the fuzzy subsets A_0, A_n and A_i $(i = 1, 2, \ldots, n-1)$ of $[\underline{a}, \overline{a}]$ by

$$\mu_{A_0}(x) = \begin{cases} f_1(x) & \text{if } x \in [a_o, a_1), \\ 0 & \text{otherwise,} \end{cases}$$

$$\mu_{A_n}(x) = \begin{cases} 1 - f_n(x) & \text{if } x \in [a_{n-1}, a_n], \\ 0 & \text{otherwise,} \end{cases}$$

$$\mu_{A_i}(x) = \begin{cases} 1 - f_i(x) & \text{if } x \in [a_{i-1}, a_i), \\ f_{i+1}(x) & \text{if } x \in [a_i, a_{i+1}), \\ 0 & \text{otherwise,} \end{cases}$$

and the numbers $u_0, u_2, \ldots, u_n \in \mathbf{R}$ by

$$u_i = f(a_i).$$

Observe that each $x \in [\underline{a}, \overline{a}]$ has a non-zero degree of membership in at most two of the fuzzy sets $A_0, A_1, \ldots, A_n$, and that we have $\sum_{i=0}^{n} \mu_{A_i}(x) = 1$. Now take an arbitrary $x \in [\underline{a}, \overline{a}]$, say, $x \in [a_{i_0-1}, a_{i_0})$ and compute the value of the input-output function induced by the Sugeno controller, yielding

$$\begin{aligned} F_S(x) &= \frac{\sum \mu_{A_i}(x) \cdot u_i}{\sum \mu_{A_i}(x)} \\ &= f(a_{i_0-1}) \cdot f_{i_0}(x) + f(a_{i_0}) \cdot (1 - f_{i_0}(x)) \\ &= f(x). \end{aligned}$$

Since x was chosen arbitrarily, this means $F_S = f$. □

The proof of Theorem 3 heavily relies on the fact that, given two different points in the two-dimensional plane, there is always exactly one straight line segment joining them. This immediately makes clear that this result does not generalize in a straightforward way to inputs of dimension two: if four points in the three-dimensional space are given, then, in general, there is no plane containing all of them. This makes it necessary to replace the interpolation in Theorem 3 by an approximation, if the inputs have dimension two or higher (see Theorem 5).

An alternative way of constructing an interpolating Sugeno controller for continuous functions, mapping a suitable subspace of $\mathbf{R}^m$ into $\mathbf{R}$, is given as follows [1]:

Theorem 4 *Let D be a compact subset of $\mathbf{R}^m$ and $f : D \to \mathbf{R}$ a continuous function. Then there exist numbers $u_1, u_2 \in \mathbf{R}$, normalized fuzzy subsets A_1, A_2 of D, and a rulebase*

$$\begin{array}{llll} \text{IF} & \mathbf{x} \textit{ is } A_1 & \text{THEN} & u = u_1 \\ \text{IF} & \mathbf{x} \textit{ is } A_2 & \text{THEN} & u = u_2 \end{array}$$

such that the input-output function F_S of the corresponding Sugeno controller coincides with f.

PROOF. Denote $u_1 = \min\{f(\mathbf{x}) \mid \mathbf{x} \in D\}$, $u_2 = \max\{f(\mathbf{x}) \mid \mathbf{x} \in D\}$, which exist because of the continuity of f and the compactness of D, define the fuzzy subsets U_1, U_2 of $[u_1, u_2]$ by

$$\begin{aligned} \mu_{U_1}(u) &= \frac{u_2 - u}{u_2 - u_1}, \\ \mu_{U_2}(u) &= \frac{u - u_1}{u_2 - u_1}, \end{aligned}$$

and introduce the fuzzy subsets A_1, A_2 of D by $A_i = f^{-1}(U_i)$ ($i = 1, 2$), i.e.,

$$\mu_{A_i}(\mathbf{x}) = \mu_{U_i} \circ f(\mathbf{x}) = \mu_{U_i}(f(\mathbf{x})).$$

Then obviously for each $u \in [u_1, u_2]$ we have $\mu_{U_1}(u) + \mu_{U_2}(u) = 1$ and, consequently, for each $\mathbf{x} \in D$ we get $\mu_{A_1}(\mathbf{x}) + \mu_{A_2}(\mathbf{x}) = 1$. Therefore, the input–output function F_S of the Sugeno controller yields

$$\begin{aligned} F_S(\mathbf{x}) &= \frac{\mu_{A_1}(\mathbf{x}) \cdot u_1 + \mu_{A_2}(\mathbf{x}) \cdot u_2}{\mu_{A_1}(x) + \mu_{A_2}(x)} \\ &= \frac{u_2 - f(\mathbf{x})}{u_2 - u_1} \cdot u_1 + \frac{f(\mathbf{x}) - u_1}{u_2 - u_1} \cdot u_2 \\ &= f(\mathbf{x}), \end{aligned}$$

which means that F_S coincides with f on the whole domain D. □

In the case of a surjective function (which can always be achieved if we define $C = f(D)$ and consider $f : D \to C$ rather than $f : D \to \mathbf{R}$) the interrelation between a fuzzy subset A of D, the corresponding U of C and the function f in Theorem 4 is given by $U = f(A)$, where $f(A)$ is defined according to the *extension principle* by

$$\mu_{f(A)} = \sup \left\{ \mu_A(\mathbf{x}) \mid \mathbf{x} \in D, u = f(\mathbf{x}) \right\}. \tag{7}$$

In the case of a one–dimensional input, i.e., if $D = [\underline{x}, \overline{x}]$ is a closed interval in $\mathbf{R}$ and f is a monotone surjective function, i.e., $f : [\underline{x}, \overline{x}] \to [\underline{u}, \overline{u}]$, then the situation in Theorem 4 corresponds to the *gradual rules* considered in [4], since for each $x \in [\underline{x}, \overline{x}]$ we have

$$\mu_{f(A_i)}(f(x)) \geq \mu_{A_i}(x). \tag{8}$$

Although the result presented in Theorem 4 applies to inputs of any finite dimension, the fuzzy subsets A_i of $\mathbf{R}^m$, in general, are not Cartesian products of suitable fuzzy subsets of $\mathbf{R}$. This can be seen if, e.g., we consider the input–output function $f : [0, 2]^2 \to [0, 1]$ given by

$$f(x, y) = e^{-(x^2 + y^2 - 1)^2}.$$

The algorithm in the proof of Theorem 4 then yields a fuzzy subset A_2 of $[0, 2]^2$ which obviously is no Cartesian product of fuzzy subsets of $[0, 2]$.

4 Approximation by fuzzy systems

As mentioned before, a straightforward generalization of the results of Theorem 3 to inputs of dimension higher than one, while keeping Cartesian products in the input space, is not possible. The following result shows that, in the two-dimensional case, all functions can be approximated by a Sugeno controller with arbitrary precision. The generalization to inputs of higher dimension is straightforward.

Theorem 5 *Let* $f : [\underline{a}, \overline{a}] \times [\underline{b}, \overline{b}] \to \mathbf{R}$ *be a function and let*

$$\begin{aligned} \underline{a} = a_0 < a_1 < \cdots < a_{m-1} < a_m = \overline{a}, \\ \underline{b} = b_0 < b_1 < \cdots < b_{n-1} < b_n = \overline{b}, \end{aligned}$$

be partitions of the intervals $[\underline{a}, \overline{a}]$ *and* $[\underline{b}, \overline{b}]$, *respectively. Then there are normalized fuzzy subsets* $A_0, A_1, \ldots, A_m$ *and* $B_0, B_1, \ldots, B_n$ *of* $[\underline{a}, \overline{a}]$ *and* $[\underline{b}, \overline{b}]$, *respectively, numbers* $u_{00}, u_{10}, \ldots, u_{mn} \in \mathbf{R}$, *and a rulebase* $(i = 0, 1, \ldots, m; j = 0, 1, \ldots, n)$

$$\text{IF} \quad x \textit{ is } A_i \quad \text{AND} \quad y \textit{ is } B_j \quad \text{THEN} \quad u = u_{ij}$$

such that the input–output function F_S *of the corresponding Sugeno controller coincides with* f *on all lattice points* (a_i, b_j), *i.e., for all* $i = 0, 1, \ldots, m$ *and all* $j = 0, 1, \ldots, n$ *we have*

$$F_S(a_i, b_j) = f(a_i, b_j). \tag{9}$$

PROOF. Define first for $i = 1, 2, \ldots, m$ and $j = 1, 2, \ldots, n$ the numbers u_{ij} by

$$u_{ij} = f(a_i, b_j),$$

the fuzzy sets A_0, A_m and A_iJ $(i = 1, 2, \ldots, m-1)$ by

$$\begin{aligned}
\mu_{A_0}(x) &= \begin{cases} \dfrac{x - a_1}{a_0 - a_1} & \text{if } x \in [a_0, a_1), \\ 0 & \text{otherwise,} \end{cases} \\
\mu_{A_m}(x) &= \begin{cases} \dfrac{x - a_{m-1}}{a_m - a_{m-1}} & \text{if } x \in [a_{m-1}, a_m], \\ 0 & \text{otherwise,} \end{cases} \\
\mu_{A_i}(x) &= \begin{cases} \dfrac{x - a_{i-1}}{a_i - a_{i-1}} & \text{if } x \in [a_{i-1}, a_i), \\ \dfrac{x - a_{i+1}}{a_i - a_{i+1}} & \text{if } x \in [a_i, a_{i+1}), \\ 0 & \text{otherwise} \end{cases}
\end{aligned}$$

and the fuzzy sets B_0, B_n and B_j $(j = 1, 2, \ldots, n-1)$ by

$$\mu_{B_0}(x) = \begin{cases} \dfrac{x - b_1}{b_0 - b_1} & \text{if } x \in [b_0, b_1), \\ 0 & \text{otherwise,} \end{cases}$$

$$\mu_{B_n}(x) = \begin{cases} \dfrac{x - b_{n-1}}{b_n - b_{n-1}} & \text{if } x \in [b_{n-1}, b_n], \\ 0 & \text{otherwise,} \end{cases}$$

$$\mu_{B_j}(x) = \begin{cases} \dfrac{x - b_{j-1}}{b_j - b_{j-1}} & \text{if } x \in [b_{j-1}, b_j), \\ \dfrac{x - b_{j+1}}{b_j - b_{j+1}} & \text{if } x \in [b_j, b_{j+1}), \\ 0 & \text{otherwise.} \end{cases}$$

Next consider the rulebase $(i = 0, 1, \ldots, m;\ j = 0, 1, \ldots, n)$

$$\text{IF} \quad x \text{ is } A_i \quad \text{AND} \quad y \text{ is } B_j \quad \text{THEN} \quad u = u_{ij}$$

and choose an arbitrary lattice point (a_{i_0}, b_{j_0}) $(i_0 = 0, 1, \ldots, m;\ j_0 = 0, 1, \ldots, n)$. Obviously we have

$$\mu_{A_{i_0} \times B_{j_0}}(a_{i_0}, b_{j_0}) = 1$$

and for $(i, j) \neq (i_0, j_0)$

$$\mu_{A_i \times B_j}(a_{i_0}, b_{j_0}) = 0,$$

and, therefore, for the input–output function of the corresponding Sugeno controller F_S we get

$$\begin{aligned} F_S(a_{i_0}, b_{j_0}) &= \frac{\sum_{i,j} \mu_{A_i \times B_j}(a_{i_0}, b_{j_0}) \cdot u_{ij}}{\sum_{i,j} \mu_{A_i \times B_j}(a_{i_0}, b_{j_0})} \\ &= \mu_{A_{i_0} \times B_{j_0}}(a_{i_0}, b_{j_0}) \cdot u_{i_0 j_0} \\ &= f(a_{i_0}, b_{j_0}). \end{aligned}$$

Since the lattice point (a_{i_0}, b_{j_0}) was chosen arbitrarily, this shows that F_S coincides with f on all lattice points (a_i, b_j). □

Remark 1 This theorem has interesting practical consequences:

(a) It is not difficult to see that the results of Theorem 5 can be generalized to arbitrary n–dimensional inputs.

(b) For each $a_i \in [\underline{a}, \overline{a}]$, $i = 0, 1, \ldots, m$, the one-dimensional section $F_S(a_i, \cdot) : [\underline{b}, \overline{b}] \to \mathbf{R}$ of the input-output function F_S of the approximating Sugeno controller in Theorem 5 is a piecewise linear function. In particular, if P_{ij} denotes the point $(a_i, b_j, f(a_i, b_j)) \in \mathbf{R}^3$, and if the four points $P_{i_0 j_0}$, P_{i_0+1, j_0}, P_{i_0, j_0+1} and P_{i_0+1, j_0+1} lie in a plane, then the restriction of the function F_S to the rectangle $[a_{i_0}, a_{i_0+1}] \times [b_{j_0}, b_{j_0+1}]$ is a linear function. This means that a classical linear controller is a special case of the Sugeno controller.

References

[1] P. Bauer, E. P. Klement, A. Leikermoser, B. Moser: Approximation of real functions by rule bases. Proceedings Fifth IFSA World Congress '93, Seoul, pp. 239–241 (1993)

[2] J. J. Buckley: Sugeno type controllers are universal controllers. Fuzzy Sets and Systems 53: 299–304 (1993)

[3] D. Driankov, H. Hellendoorn, M. Reinfrank: An Introduction to Fuzzy Control. Springer, Berlin (1993)

[4] D. Dubois, M. Grabisch, H. Prade: Gradual rules and the approximation of functions. Proceedings of the 2nd International Conference on Fuzzy Logic and Neural Networks, Iizuka, pp. 629–632 (1992)

[5] B. Kosko: Fuzzy systems as universal approximators. Proceedings FUZZ-IEEE '92, San Diego, pp. 1153–1162. IEEE (1992)

[6] J. Lee, S. Chae: Analysis on function duplicating capabilities of fuzzy controllers. Fuzzy Sets and Systems 56: 127–143 (1993)

[7] E. H. Mamdani, B. R. Gaines: Fuzzy Reasoning and Its Applications. Academic Press, New York (1981)

[8] M. Sugeno: An introductory survey of fuzzy control. Inform. Sci. 36: 59–83 (1985)

[9] L. X. Wang: Fuzzy systems are universal approximators. Proceedings FUZZ-IEEE '92, San Diego, pp. 1163–1169. IEEE (1992)

5.5. Defuzzification As Crisp Decision Under Fuzzy Constraints — New Aspects of Theory and Improved Defuzzification Algorithms

Thomas A. Runkler and Manfred Glesner

Abstract

Although defuzzification is an essential functional part of all fuzzy systems, it is not firmly embedded in fuzzy theory yet.

Proceeding from fuzzy decision theory we define defuzzification as crisp decision under fuzzy constraints and achieve a new theoretical foundation of the defuzzification process. From these theoretical considerations we develop a new class of lucidly customizable defuzzification procedures (constrained decision defuzzification *CDD*). We develop powerful examples of CDD customizations and show that CDD is superior to the standard defuzzification algorithms center of gravity and mean of maxima, represented by the parametric basic defuzzification distribution (*BADD*), concerning static, dynamic and statistical properties.

1 Introduction

Since the concept of fuzzy sets was first introduced by [Zadeh, 1965], it has been applied to many different fields, ranging from engineering applications (control [Runkler *et al.*, 1992, e.g.], robotics, image and speech processing), biological and medical sciences to applied operations research and expert systems [Zimmermann, 1987].

Fuzzy systems internally process fuzzy values, which have to be mapped to crisp output in most applications. This mapping is called **defuzzification**. Although defuzzification is an essential part of most fuzzy systems, it is not firmly embedded in fuzzy set theory yet.

We show that the results of fuzzy decision theory [Dubois and Prade, 1980, Zimmermann, 1987] serve as a theoretical foundation of defuzzification. We develop a new theoretical approach defining defuzzification as *crisp decision under fuzzy constraints* (sections 2 and 3).

The theoretical results achieved are of high practical relevance, because they lead to a new class of lucidly customizable defuzzification procedures we refer to as *constrained decision defuzzification (CDD)*. In section 4 we develop examples of some rational customizations of CDD.

There are several categories of defuzzification method properties: static, dynamic and statistical. Concerning these properties we examine the new CDD method in comparison with the standard methods *center of gravity (COG)* and *mean of maxima (MOM)*. Even more generally we compare CDD with the *basic defuzzification distribution (BADD)* method [Filev and Yager, 1991, Yager, 1992], a parametric procedure which includes COG and MOM as special cases (section 5).

2 Decisions Under Fuzzy Constraints

A *decision* is defined as the solution of a given multicriteria optimization problem with goals $\tilde{G}_j$ and constraints $\tilde{C}_i$. In a fuzzy environment, we represent goals and constraints as fuzzy sets formally having the same nature. A *fuzzy decision* [Dubois and Prade, 1980, Zimmermann, 1987] (see Fig. 1) is defined as

$$\tilde{D}(\tilde{G}_1, \ldots, \tilde{G}_m, \tilde{C}_1, \ldots, \tilde{C}_n) := \bigcap_{j=1,\ldots,m} \tilde{G}_j \cap \bigcap_{i=1,\ldots,n} \tilde{C}_i. \quad (1)$$

Without loss of generality we assume decisions with only one goal, hence

$$\tilde{D}(\tilde{G}, \tilde{C}_1, \ldots, \tilde{C}_n) = \tilde{G} \cap \bigcap_{i=1,\ldots,n} \tilde{C}_i. \quad (2)$$

We define the *crisp decision* $d(\tilde{G}, \tilde{C}_1, \ldots, \tilde{C}_n)$ under fuzzy constraints (see Fig. 1) as the mean value[1] of all $x \in \mathcal{X}$ with maximum membership in $\tilde{D}(\tilde{G}, \tilde{C}_1, \ldots, \tilde{C}_n)$.

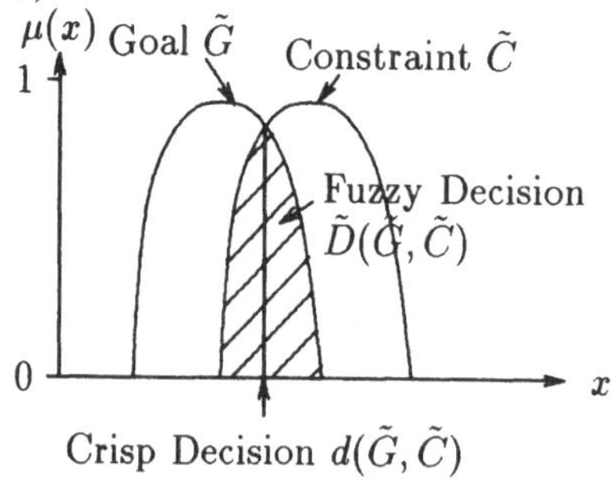

Fig. 1: Fuzzy and crisp decision

[1] Other realizations are possible, but for optimization problems the mean of maxima value is the most reasonable.

Example

The goal is to determine a desired room temperature ϑ close to $20°C$, which satisfies the following constraints:

1. One inhabitant is ill, so the temperature should be rather high, and,

2. to avoid frost damage, the temperature must be higher than $0°C$.

Solution

We choose membership functions for the goal and the constraints as

$$\mu_{\tilde{G}}(\vartheta) = \begin{cases} 0 & \text{for } \vartheta \leq 10°C, \\ \frac{\vartheta - 10°C}{10K} & \text{for } 10°C < \vartheta \leq 20°C, \\ 1 - \frac{\vartheta - 20°C}{10K} & \text{for } 20°C < \vartheta \leq 30°C, \\ 0 & \text{for } \vartheta > 30°C, \end{cases} \tag{3}$$

$$\mu_{\tilde{C}_1}(\vartheta) = \begin{cases} 0 & \text{for } \vartheta \leq 0°C, \\ \frac{\vartheta - 0°C}{40K} & \text{for } 0°C < \vartheta \leq 40°C, \\ 1 & \text{for } \vartheta > 40°C, \end{cases} \tag{4}$$

$$\mu_{\tilde{C}_2}(\vartheta) = \begin{cases} 0 & \text{for } \vartheta \leq 0°C, \\ 1 & \text{for } \vartheta > 0°C. \end{cases} \tag{5}$$

As shown in Fig. 2, we determine the crisp decision, i.e. the desired temperature as $\vartheta = 24°C$.

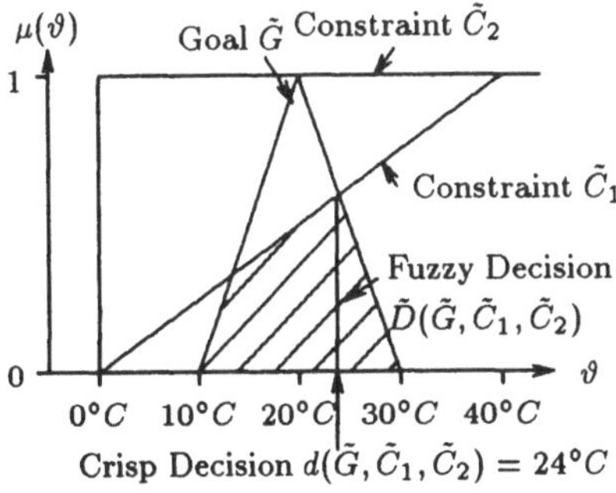

Fig. 2: Example: temperature decision

3 Defuzzification as Decision

3.1 The Mean Of Maxima Method

The *mean of maxima* defuzzification can be interpreted as a **crisp decision without constraint**[2], where $\tilde{Q}$, the **fuzzy system output, serves as the goal**. We express defuzzification by a **defuzzification operator** $\mathcal{F}^{-1}$ mapping membership functions to crisp numbers and write:

$$\mathcal{F}_{\text{MOM}}^{-1}\left(\mu_{\tilde{Q}}(x)\right) = d(\tilde{Q}, \mathcal{X}). \tag{6}$$

3.2 Extended Defuzzification as Decision Under Non Trivial Constraints

In general we allow non trivial constraints for the decision making process and achieve a formula for the *constrained decision defuzzification (CDD)*:

$$\mathcal{F}_{\tilde{C}_1,\ldots,\tilde{C}_n}^{-1}\left(\mu_{\tilde{Q}}(x)\right) = d(\tilde{Q}, \tilde{C}_1, \ldots, \tilde{C}_n). \tag{7}$$

Generally, the constraints $\tilde{C}_i$ depend on the fuzzy system output $\tilde{Q}$:

$$\tilde{C}_i = \tilde{f}_i(\tilde{Q}). \tag{8}$$

The constrained defuzzification operator (7) is a general purpose defuzzification method. CDD is easily customizable using lucid constraint definitions. In the following, we give two typical examples for rational definitions of constraints.

4 Examples

4.1 Constraint "Near Center"

Typically, the defuzzified value is required to be near the center of the membership function area, i.e. the product of the defuzzified value and the membership function area should be similar to the area's first moment:

$$\mu_{\tilde{C}_1}(x) := \text{sim}\left(x \int_{\mathcal{X}} \mu_{\tilde{Q}}(\xi)\, d\xi, \int_{\mathcal{X}} \xi \cdot \mu_{\tilde{Q}}(\xi)\, d\xi\right). \tag{9}$$

[2]Formally, this is achieved using the universal set $\mathcal{X}$ as constraint.

We choose the gaussian function as a "natural" similarity function:

$$\operatorname{sim}(a,b) := e^{-\frac{(a-b)^2}{2\sigma^2}} \in (0,1]. \tag{10}$$

If we choose the standard deviation $\sigma \to \infty$, then $\tilde{C}_1 \to \mathcal{X}$, and we achieve the *mean of maxima method.* If we choose the standard deviation $\sigma \to 0$, then

$$\mu_{\tilde{C}_1}(x) = \begin{cases} 1 & \text{if } x \int\limits_{\mathcal{X}} \mu_{\tilde{Q}}(\xi)\, d\xi = \int\limits_{\mathcal{X}} \xi \cdot \mu_{\tilde{Q}}(\xi)\, d\xi \\ 0 & \text{else,} \end{cases} \tag{11}$$

i.e. $\tilde{C}_1$ becomes a singleton at the center of gravity of $\tilde{Q}$. Thus, we achieve the *center of gravity (COG)* method for $\sigma \to 0$.

4.2 Constraint "Membership $\geq \alpha$"

In many applications the membership of the defuzzified value is required to be larger than a certain value α. This can be forced by the constraint (see Fig. 3)

$$\mu_{\tilde{C}_2}(x) = \begin{cases} \frac{\mu_{\tilde{Q}}(x)-\alpha}{1-\alpha} & \text{if } \mu_{\tilde{Q}}(x) > \alpha \\ 0 & \text{else.} \end{cases} \tag{12}$$

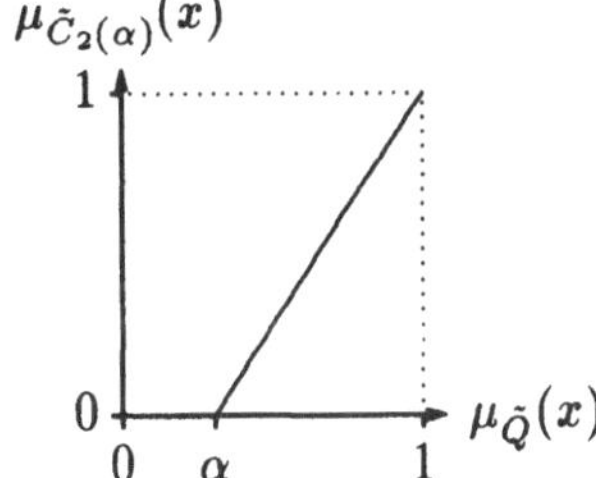

Fig. 3: Constraint "membership $\geq \alpha$"

5 Properties of the Constrained Defuzzification

We consider the CDD method $\mathcal{F}^{-1}_{\tilde{C}_1,\tilde{C}_2}(\mu_{\tilde{Q}}(x))$ using (9), (12) and examine its properties in comparison with the *basic defuzzification distribution (BADD) method* [Filev and Yager, 1991, Yager, 1992]. BADD is a parametric defuzzification method including the standard methods COG and

MOM as special cases. The BADD defuzzification is defined as:

$$\mathcal{F}^{-1}_{\text{BADD}}\left(\mu\left(x\right),\gamma\right) := \frac{\int_{x_{\inf}}^{x_{\sup}} \mu(x)^{\gamma} \cdot x \, dx}{\int_{x_{\inf}}^{x_{\sup}} \mu(x)^{\gamma} \, dx} \tag{13}$$

with

$$\mathcal{F}^{-1}_{\text{BADD}}\left(\mu\left(x\right),\gamma=1\right) = \mathcal{F}^{-1}_{\text{COG}}\left(\mu\left(x\right)\right), \tag{14}$$

$$\mathcal{F}^{-1}_{\text{BADD}}\left(\mu\left(x\right),\gamma\rightarrow\infty\right) = \mathcal{F}^{-1}_{\text{MOM}}\left(\mu\left(x\right)\right). \tag{15}$$

5.1 Static Properties

[Runkler and Glesner, 1993c] developed a set of rational constraints for defuzzification methods. We apply these constraints to CDD to examine its static properties.

As well as BADD, the CDD method violates the following rational constraints: strong μ–translation, t–norm, fuzzy number and prohibitive information property. Additionally, the μ–scaling property is violated. This is due to the fact, that the threshold value α remains constant when μ is scaled. This violation is of minor importance, i.e., concerning the static properties, CDD performs like BADD.

5.2 Dynamic Properties

We assume the typical case of a knowledge base with two symmetric, triangular, connected output membership functions **"low"** and **"high"**, with firing grades h and $1-h$ (see Fig. 4)[3] [Runkler and Glesner, 1993b]. For changing values of the firing grade[4] h (i.e. for changing input) we calculate the defuzzified output using BADD and CDD for different parameter values, and obtain Fig. 5.

For changing values of γ the BADD output steadily slides from the mean of "high" to the mean of "low" with different slopes. The CDD output behaves differently: For small values of σ it slides from the mean of "high" to the mean of "low", while for increasing values of σ "low μ zones" are neglected.

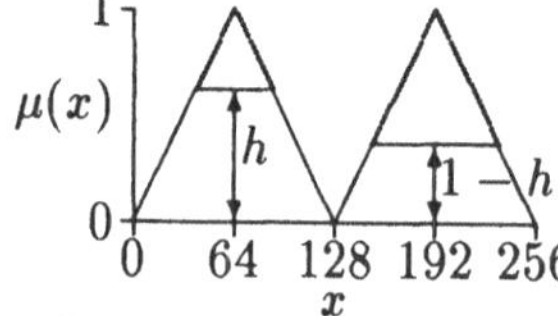

Fig. 4: Connected membership functions

[3] The output range [0, 255] is chosen for convenience only.

[4] We assume max–min–inference.

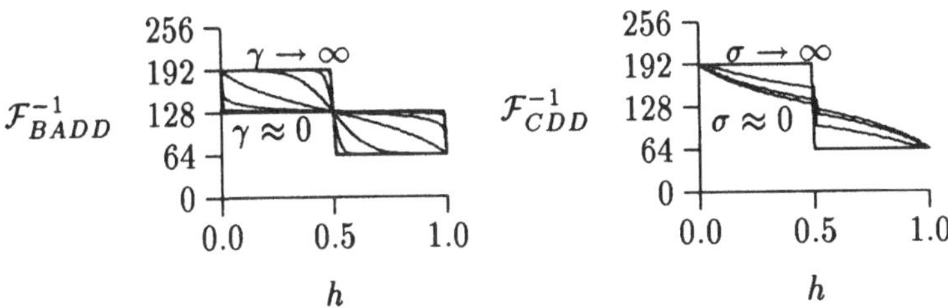

Fig. 5: Dynamic output using connected membership functions

We examine the case of separate membership functions as shown in Fig. 6. BADD behaves like before: The output slides from "high" to "low" crossing the **forbidden area** where $\mu = 0$ (marked with dashed lines). This behavior is not acceptable in most applications (e.g. robotics [Pfluger *et al.*, 1992]). CDD shows a more desirable behavior: It strictly avoids zero μ zones (cut off) and prefers high μ zones (see Fig. 7).

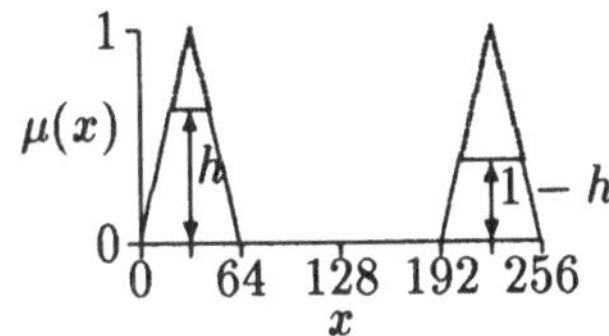

Fig. 6: Separate membership functions

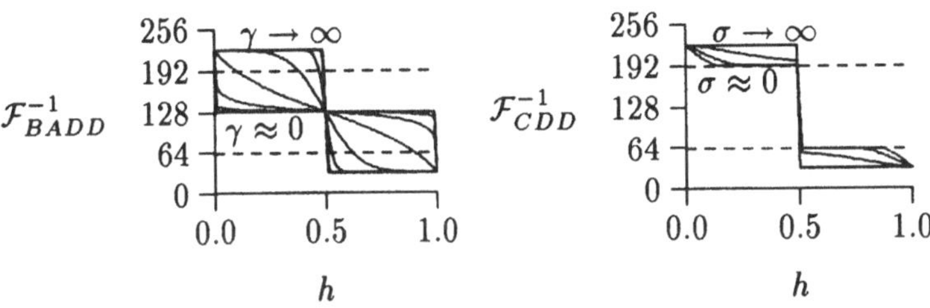

Fig. 7: Dynamic output using separate membership functions

We have seen, that BADD defuzzification leads to undesirable results, when zero μ intervals are used, while CDD prefers output with high membership leading to more reasonable results.

5.3 Statistical Properties

We measure BADD and CDD's preference for large membership values by defuzzifying random membership functions. We calculate the membership frequencies of the defuzzified values, $h\left(\mu\left(\mathcal{F}^{-1}(\mu(x))\right)\right)$, and achieve Fig. 8.

BADD results in output with the average membership 0.5 to 1. Even for high belief (high γ) it frequently selects output with very low membership. CDD prefers output with membership between 0.65 and 1. Low μ output is rare, and output with membership less than the threshold α is strictly avoided.

We have shown, that from the statistical point of view, CDD is superior to BADD, because it prefers values with high membership.

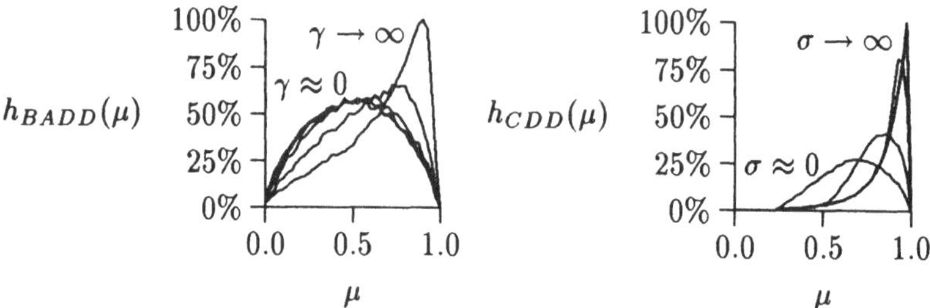

Fig. 8: Membership frequencies

5.4 Implementation Effort

The effort of the implementation of the CDD method strongly depends on the constraints and the hardware used. For the CDD example regarded in this paper the calculation effort is of the same dimension as for COG, and lower than for other BADD methods.

If the fuzzy system is realized *approximately*, e.g. using B–splines [Runkler and Glesner, 1993a], the CDD operations can even be executed at compile time, decreasing the run time effort. Therefore the use of this advanced defuzzification method does not necessarily increase the system run time.

6 Results

We developed a new theoretical approach for defuzzification based on the fuzzy decision theory. We have shown that defuzzification can be regarded as a crisp decision under fuzzy constraints.

These theoretical examinations resulted in the definition of a new class of lucidly customizable defuzzification procedures referred to as *constrained decision defuzzification (CDD)*. We developed a typical customization example.

We regarded three different categories of properties of defuzzification operators: static, dynamic and statistical aspects. Compared to the parametric BADD method, the typically customized CDD defuzzification performed excellently. Concerning the static behavior, BADD and CDD results are almost equal. The CDD dynamics are superior to BADD's, especially when separate membership functions or zones with low membership values are used. Also concerning the membership values statistics, CDD behaves much better than BADD.

Constrained decision defuzzification is a step towards a firm theoretical foundation of defuzzification and towards the design of new and better defuzzification operators.

Acknowledgments

This work is a part of the doctoral research programme of T. A. Runkler within the Graduiertenkolleg "Intelligent Systems for Information and Automation Technology" at Darmstadt University of Technology.

In this novel programme, PhD students (electronic engineers and computer scientists) are working together on a wide variety of problems concerning theory and realization of intelligent systems and modern methods of information processing.

Mailing Address

Thomas A. Runkler, Darmstadt University of Technology, Institute of Microelectronic Systems, Karlstraße 15, 64283 Darmstadt, Tel. (+49) 6151 16 - 4337, Fax (+49) 6151 16 - 4936, Email *thomasr@microelectronic.e-technik.th-darmstadt.de*.

References

[Bandemer and Gottwald, 1992] H. Bandemer and S. Gottwald. *Einführung in Fuzzy Methoden*. Akademie Verlag, Berlin, third edition, 1992.

[Dubois and Prade, 1980] D. Dubois and H. Prade. *Fuzzy Sets and Systems*. Academic Press, London, 1980.

[Filev and Yager, 1991] D. P. Filev and R. R. Yager. A generalized defuzzification method via bad distributions. *International Journal of Intelligent Systems*, 6:687-697, 1991.

[Pfluger *et al.*, 1992] N. Pfluger, J. Yen, and R. Langari. A defuzzification strategy for a fuzzy logic controller employing prohibitive information in command formulation. In *IEEE International Conference on Fuzzy Systems*, pages 717–723, 1992.

[Runkler and Glesner, 1993a] T. A. Runkler and M. Glesner. Approximative Synthese von Fuzzy-Controllern. In B. Reusch, editor, *Fuzzy Logic — Theorie und Praxis*, pages 22–31. Springer, Berlin, 1993.

[Runkler and Glesner, 1993b] T. A. Runkler and M. Glesner. Defuzzification with improved static and dynamic behavior: Extended center of area. In *European Congress on Fuzzy and Intelligent Technologies*, pages 845–851, Aachen, September 1993.

[Runkler and Glesner, 1993c] T. A. Runkler and M. Glesner. A set of axioms for defuzzification strategies — towards a theory of rational defuzzification operators. In *IEEE International Conference on Fuzzy Systems, San Francisco*, pages 1161–1166, March/April 1993.

[Runkler *et al.*, 1992] T. A. Runkler, H.-J. Herpel, and M. Glesner. Einsatz von Fuzzy Logic in einer intelligenten Kupplung. In *VDE-Fachtagung "Technische Anwendungen von Fuzzy-Systemen"*, pages 120–126, Dortmund, November 1992.

[Yager, 1992] R. R. Yager. Fuzzy sets and approximate reasoning in decision and control. In *IEEE International Conference on Fuzzy Systems*, pages 418–428, 1992.

[Zadeh, 1965] L. A. Zadeh. Fuzzy sets. In *Information and Control 8*, pages 338–353, 1965.

[Zimmermann, 1987] H. J. Zimmermann. *Fuzzy Sets, Decision Making, and Expert Systems*. Kluwer Academic Publishers, Boston, 1987.

5.6. Uncertainty and Fuzziness

Wolfgang Sander

Abstract

There are different mathematical frameworks dealing with uncertainty, vagueness and ambiguity : the probabilistic concept, the concept of a fuzzy set, and the concept of a fuzzy measure. The corresponding measures for the amount of relevant information lead to three types of uncertainty measures : Entropies or measures of information, measures of fuzziness and the uncertainty measures in the mathematical theory of evidence.

One purpose of this paper is to focus on recent results of measures of fuzziness and to give a survey on characterizations of these measures. Moreover, we want to show that certain "total entropies" which consist of a "random part" and a "fuzzy part", are special cases of a more general information theory, where the entropies are dependent upon the events and the probabilities.

1 Introduction

We first remember some basic definitions. In this paper we consider only discrete, finite sets $\Omega = \{x_1, \ldots, x_n\}$. If $f, g : \Omega \to [0,1]$ and $h : \Omega' \to [0,1]$ are fuzzy sets, then $f \vee g, f \wedge g \in [0,1]^\Omega$ and $f \times h \in [0,1]^{\Omega \times \Omega'}$ are the functions defined by

$$(f \vee g)(x) = max\{f(x), g(x)\}, (f \wedge g)(x) = min\{f(x), g(x)\}$$

and

$$(f \times g)(x,y) = f(x)g(y)$$

for all $x \in \Omega$ and $y \in \Omega'$; moreover,

$$P(f) = \sum_{i=1}^{n} f(x_i)$$

is the power of f.

Now we assign a nonnegative real number to each fuzzy set $f \in [0,1]^\Omega$ that characterizes the degree of fuzziness of f.

Definition 1

If $\Omega = \{x_1, \ldots, x_n\}$ is a finite set, then every function $d : [0,1]^\Omega \to \mathbb{R}_o = [0, \infty)$ satisfying

$$d(f) = I_n(f(x_1), \ldots, f(x_n)), \qquad f \in [0,1]^\Omega,$$

for some mapping $I_n : [0,1] \to \mathbb{R}_o$ is called a measure of fuzziness or a fuzzy entropy.

This definition says that a fuzzy entropy can be regarded as an entropy in the sense that it measures the uncertainty about the presence or absence of a certain property over the set Ω. In the case of a finite basic set Ω the fuzzy entropy $d(f)$ is determined by the function values of Ω.

The above definition is analogous to the following definition of more general (probabilistic) entropies. If we interpret the elements $x_1, \ldots, x_n$ of Ω as elements of a ring B of subsets (containing, with any two sets,their union and difference, and thus also their intersection and the empty set $\emptyset$) and as the possible events of an experiment having probabilities $p_1, \ldots, p_n$, respectively then we get the following definition using the notations

$$\Gamma_n = \{P = (p_1, \ldots, p_n) : \sum_{i=1}^{n} p_i = 1, p_i \geq 0\}, \quad n \geq 2 \tag{1}$$

and

$$\Omega_n = \{X = (x_, \ldots, x_n) : x_i \epsilon B, x_i \cap x_j = \emptyset, i \neq j\}, \quad n \geq 2 : \tag{2}$$

Definition 2

Let $\Omega = \{x_1, \ldots, x_n\}$ be a finite set. Then every sequence (J_n) or (G_n) with

$$J_n : \Gamma_n \to \mathbb{R} \qquad or \qquad G_n : \Omega_n \times \Gamma_n \to \mathbb{R}$$

is called an entropy or a generalized entropy (inset entropy fo short) , respectively.

We remark that generalized entropies are also called inset information measures (the name was chosen because the idea for these type of measures was born at a meeting at the École Normale Supérieure de l'Enseignment Technique - ENSET - near Paris). Note that the measures (G_n) may depend on both the probabilities and the events [Aczél, Daróczy 1978] whereas (J_n) depend only upon the probabilities.

2 Characterizations of fuzzy entropies

A main purpose of an axiomatic theory for measures of uncertainty is to find and characterize measures of uncertainty having certain useful properties. Concerning entropies and generalized entropies there are a lot of characterization theorems [Aczél, Daróczy 1975, Aczél 1984, Aczél, Daróczy 1978, Sander 1987, Groß 1993] whereas in the mathematical theory of evidence there are only few results in this direction [Klir, Folger 1988, p. 297ff]. In this section we want to focus on measures of fuzziness or fuzzy entropies. There are many "useful" properties to qualify as a meaningful fuzzy entropy [Dubois, Prade 1980, p.32ff, Klir, Folger 1988, p. 140ff] and we can find many proposals for measures of fuzziness. But in our opinion there are missing characterization theorems. Thus we want to point out how to get characterization theorems using analogies with characterization theorems for (probabilistic) entropies. The following results can be found in [Ebanks 1983] and [Sander 1989].

Let us start with five well-known natural properties of fuzzy entropies [De Luca, Termini 1972]:

(P1) Sharpness : $d(f) = 0$ iff $f(\Omega) \subset \{0, 1\}$.
(P2) Maximum : $d(f)$ maximum iff $f(\Omega) = \{\frac{1}{2}\}$.
(P3) Sharpness Relation : Let $f, g \in [0, 1]^\Omega$.

$$If\ f \prec g\ then\ d(f) \leq d(g).$$

(P4) Symmetry : $d(f) = d(1 - f)$ for all $f \in [0, 1]^\Omega$.
(P5) Valuation : $d(f \vee g) + d(f \wedge g) = d(f) + d(g)$ for all $f, g \in [0, 1]^\Omega$.
The relation $f \prec g$ in (P3) is defined by

$$g(x) \geq f(x) \quad for \quad f(x) \geq \frac{1}{2}$$
$$g(x) \leq f(x) \quad for \quad f(x) \leq \frac{1}{2} \qquad (x \in \Omega).$$

It turns out that property (P5) plays a key role.

Theorem 1

Let $\Omega = \{x_1, \ldots, x_n\}$ and let $d : [0, 1]^\Omega \to \mathbb{R}_o$ be a fuzzy entropy. Then d satisfies (P5) iff there exist mappings $D_{n,i} : [0, 1] \to \mathbb{R}_o (1 \leq i \leq n)$ such that

$$d(f) = \sum_{i=1}^{n} D_{n,i}(f(x_i)) \qquad (3)$$

If $d(f) = I_n(f(x_1), \dots, f(x_n))$ is also symmetric in its variables then (P5) is equivalent to the existence of a generating function $D_n : [0,1] \to \mathbb{R}$ such that

$$d(f) = \sum_{i=1}^{n} D_n(f(x_i)), \qquad f \in [0,1]^{\Omega}. \tag{4}$$

In the probabilistic information theory the property (4) is analogous to the sum form property

$$J_n(P) = J_n(p_1, \dots, p_n) = \sum_{i=1}^{n} F(p_i), \qquad P \in \Gamma_n$$

with a generating function $F : [0,1] \to \mathbb{R}$.

It is clear that we can characterize all fuzzy entropies satisfying (4) and one of the properties (Pi), $1 \leq i \leq 4$. For example, d satisfies (4) and (P1) iff $D_n(0) = D_n(1) = 0$ and $D_n(u) > 0$ for all $u \in (0,1)$.

Let us here consider fuzzy sets with the additional property $P(f) = 1$. If we interpret $f(x)$ as the probability that $x \in \Omega$ possesses a certain property E, and think of the process of deciding whether $x_1, x_2, \dots, x_n$ do or do not possess E as an experiment then (P1) goes over into the (probabilistic) property of decisiveness

$$D_n(1,0,\dots,0) = \dots D_n(0,\dots,0,1,0,\dots,0) = D_n(0,0,\dots,0,1) = 0$$

with the regularity property $I_n(P) > 0$ (where $P \in \Gamma_n$, but P is not a unit vector in $\mathbb{R}^n$).

To get the explicit form of $D_{n,i}$ and D_n in (3) and (4), respectively we consider the "natural" property

$$Subadditivity : d(f \times g) \leq d(f) + d(g)$$

($f \in [0,1]^{\Omega}, g \in [0,1]^{\Omega'}$). By introducing appropriate weight factors we arrive at the following two properties:

(P6) There exist mappings $s, t : \mathbb{R}_o \to \mathbb{R}$ such that

$$d(f \times g) = t[P(g)] \cdot d(f) + s[P(f)] \cdot d(g)$$

and

(P7) There exist mappings $s, t : [0,1] \to \mathbb{R}_o$ such that

$$d(f \times g) = P[t(g)] \cdot d(f) + P[s(f)] \cdot d(g)$$

(in both properties we have $f \in [0,1]^{\Omega}$, $g \in [0,1]^{\Omega'}$, where Ω, Ω' are finite sets). We now consider a fuzzy entropy d as a sequence (I_n), so that (P6) and (P7) refer to families of measures of fuzziness

$$\{d : [0,1]^{\Omega} \to \mathbb{R}_o \mid \, |\Omega| < \aleph_o\}.$$

Using results from the theory of functional equations the following characterization theorem can be proven:

Theorem 2

A measure d of fuzziness (over sets Ω of all finite sizes) satisfies

(1) (P1),(P5) and (P6) iff

$$d(f) = -c \sum_{x \in \Omega} f(x) \cdot \ln f(x), \qquad f \in [0,1]^{\Omega}$$

where $c \in (0, \infty)$ and $0 \cdot \ln 0 := 0$.

(2) (P1),(P2),(P3),(P5) and (P7) iff d(f) has the form

$$d(f) = -c \sum_{x \in \Omega} f(x)^a \cdot \ln f(x), \qquad f \in [0,1]^{\Omega}$$

where $c \in (0, \infty)$, $a = \log_2 e$, $0 \cdot \ln 0 := 0$, or

$$d(f) = k \cdot \sum_{x \in \Omega} (f(x)^a - f(x)^b)$$

where $k \in (0, \infty)$, $a, b \in \mathbb{R}_o, a \leq b$ and $a \cdot 2^{-a} = b \cdot 2^{-b}$.

(3) (P1) to (P5) and (P7) iff

$$d(f) = k \cdot \sum_{x \in \Omega} f(x)(1 - f(x)), \qquad f \in [0,1]^{\Omega}$$

where $k \in (0, \infty)$.

The fuzzy entropy given in part (1) of Theorem 2 does not satisfy (P2) to (P4). Thus there is no fuzzy entropy fulfilling (P1) to (P6). But

$$H(f) := d(f) + d(1 - f) = \sum_{x \in \Omega} S(f(x)) \tag{5}$$

where d is given by part (1) of Theorem 2 satisfies (P1) to (P5) (Here $S : [0,1] \to \mathbb{R}$ is the Shannon function $S(u) = -u \ln u - (1-u)\ln(1-u), u \in [0,1]$).

Again, let us consider in Theorem 2 fuzzy sets f with the additional property $P(f) = 1$. Then - for example - property (P6) goes over into the additivity (since from the proof of Theorem 2 we get that s(1) = t(1) = 1):

$$d(f \times g) = d(f) + d(g).$$

Therefore part(1) of Theorem 2 is analogous to the fact that the Shannon entropy

$$H_n = -\sum_{i=1}^{n} p_i \log_2 p_i, \qquad P \in \Gamma_n$$

is determined by additivity, the sum form property and a regularity property for H_n, up to a multiplicative constant. Moreover, part (2) of Theorem 2 is similar to the result, that the entropies of degree (a,b)

$$H_n^{(a,b)} = \begin{cases} H_n(P) & if \quad a = b \\ k \cdot \sum_{i=1}^{n} (p_i^a - p_i^b) & if \quad a \neq b \end{cases}$$

($k \in \mathbb{R}$) are characterized by a generalized additivity condition, the sum form property and a regularity property for $H_n^{(a,b)}$.

3 Inset entropies

We first remember the following well-known result on inset entropies [Aczél, Daróczy 1978 , Aczél, Kannappan 1978].

Theorem 3
Let $a \in \mathbb{R}\backslash 0$ and let $G_n : \Omega_n \times \Gamma_n \to \mathbb{R}$ $(n \geq 2)$ be an inset entropy with $\Omega = \bigcup_{i=1}^{n} x_i$. Then G_n is symmetric,

$$G_n \begin{pmatrix} x_1, \ldots, x_n \\ p_1, \ldots, p_n \end{pmatrix} = G_n \begin{pmatrix} x_{\pi 1}, \ldots, x_{\pi n} \\ p_{\pi 1}, \ldots, p_{\pi n} \end{pmatrix}, \pi \qquad permutation,$$

a-recursive, that is ,

$$G_n \begin{pmatrix} x_1, \ldots, x_n \\ p_1, \ldots, p_n \end{pmatrix} = G_{n-1} \begin{pmatrix} x_1 \cup x_2, x_3, \ldots, x_n \\ p_1 + p_2, p_3, \ldots, p_n \end{pmatrix} +$$

$$+(p_1 + p_2)^a \cdot G_2 \begin{pmatrix} x_1, x_2 \\ \frac{p_1}{p_1+p_2}, \frac{p_2}{p_1+p_2} \end{pmatrix}$$

$$(n \geq 3 \quad ; \quad 0 \cdot G_2 \begin{pmatrix} x_1, x_2 \\ 0/0, 0/0 \end{pmatrix} = 0)$$

and measurable, that is,

$$p \rightarrow G_2 \begin{pmatrix} x_1, x_2 \\ 1-p, p \end{pmatrix}, \qquad p \in [0,1] \textit{ is measurable,}$$

iff G_n *has the form*

$$G_n(X,P) = cH_n^a(P) + \sum_{i=1}^{n} p_i^a g(x_i), \quad X \in \Omega_n, \quad P \in \Gamma_n \tag{6}$$

where $g : B \rightarrow \mathbb{R}$ *is an arbitrary function. Here* $H_n^1(P) = H_n(P)$ *is the Shannon entropy and*

$$H_n^a(P) = (2^{1-a} - 1)^{-1}[(\sum_{i=1}^{n} p_i^a) - 1], \qquad a \neq 1, \qquad P \in \Gamma_n$$

is the entropy of degree a (with the convention $0^a = 0$*).*

We have presented this result to show that the total entropy [De Luca, Termini 1972]

$$H_{tot}(f, P, X) = H_n(P) + \sum_{i=1}^{n} p_i S(f(x_i))$$

($P \in \Gamma_n, f \in [0,1]^\Omega, X \in \Omega_n$), which consists of a random term and a fuzzy term, has exactly the form of the above inset entropies (if a = 1)with $g = S \circ f$ (We have used the above notations). Thus we can characterize $H_{tot}(f, P, X)$ using Theorem 3.1. and the results of chapter 2 (If, in partcular, $P = (1/n, \ldots, 1/n)$ then we have

$$H_{tot}(f, P, X) - \log n = 1/n \cdot H(f)).$$

On the other hand the result in Theorem 3 gives us a total entropy for $a \neq 1$ which is analogous to H_{tot}. This means that we can present a "natural" one-parametric generalization of H_{tot} by putting $g = S \circ f$ in equation (6).

4 Concluding Remarks

In this note we wanted to point out the advantage of characterization theorems for fuzzy entropies and total entropies for applications : An user can choose an appropriate fuzzy entropy knowing some natural properties of the apriori unknown fuzzy entropy.

Because there exist far reaching generalizations of Theorem 3 [see the survey paper of Sander 1987], it is possible to introduce extensive classes of total entropies which cover also recent results in the theory of evidence. We come back to these results in another paper.

References

[Aczél 1984] J.Aczél: Measuring information beyond communication theory–Why some generalized information measures may be useful, others not.Aequationes Math. 27, 1-19 (1984)

[Aczél, Daróczy 1975] J.Aczél, Z.Daróczy: On measures of information and their characterizations. Academic Press, New York–San Francisco–London (1975)

[Aczél, Daróczy 1978] J.Aczél, Z.Daróczy: A mixed theory of information–I:Symmetric,recursive and measurable entropies of randomized systems of events. Rev. Francaise Automat. Informat. Recherche Operationelle. Ser. Rouge Inform. Theor. 12, 149–155 (1978)

[Aczél, Kannappan 1978] J.Aczél, PL.Kannappan: A mixed theory of information–III: Inset entropies of degree β, Information and Control 1978, 312-322 (1978)

[De Luca, Termini 1972] A.De Luca, S.Termini: A definition of a non-probabilistic entropy in the setting of fuzzy sets theory, Information and Control 20, 301-312 (1972)

[Dubois, Prade 1980] D.Dubois, H.Prade: Fuzzy Sets and Systems, Theory and Applications, Academic Press. New York–London 1980

[Ebanks 1983] B.R.Ebanks: On measures of fuzziness and their representations, J.Math. Anal. Appl., 24-37 (1983)

[Groß 1993] H.Groß : Inset-Informationsmaße, Dissertation. Braunschweig 1993

[Klir, Folger] G.J.Klir, T.A.Folger: Fuzzy Sets, Uncertainty, and Information, Prentice Hall,New York (1988)

[Sander 1987] W.Sander: The fundamental equation of information and its generalizations, Aequationes Math. 33, 150-182 (1987)

[Sander 1989] W.Sander: On measures of fuzziness, Fuzzy Sets and Systems 29, 49-55 (1989)

6

Fuzzy Classification

6.1. Fuzzy Classification: An Overview

Klaus Dieter Meyer Gramann

Abstract

Fuzzy classification, fuzzy diagnosis, and fuzzy data analysis are - besides fuzzy control - the most important application areas of fuzzy logic. In this chapter four practical tasks are presented which can roughly be characterized as technical classification and diagnosis, fuzzy data analysis in chemical model creation, medical object recognition, and decision making support by a life insurance. Neural networks and analytical methods of classical statistics try to find explicitely a classifying function with the help of a sample. The development of knowledge-based systems has stimulated modern constructive approaches like IF-THEN-rules and causal networks. In order to deal with vague observations, vague relationships between features, and/or non-crip classification, these analytical and constructive methods were transfered from crisp numbers to fuzzy sets. The contributions of fuzzy sets to the four applications of this chapter are presented. Some general remarks on the applicability and limitations of fuzzy classification conclude this short introduction to fuzzy classification.

1 Introduction

"Fuzzy logic" has become one buzzword of the nineties. Some technical devices with the label "fuzzy" were economically successful, especially in Japan, and are often mentioned in the fuzzy literature: traffic systems, electrical and electronical consumer appliances, automatization systems, components of automobiles, etc. Nearly all of these fuzzy applications can be called fuzzy controllers: The task is to control a closed loop automatically such that specific requirements (the control gains) are fulfilled. A chapter of this book is devoted to "some trends in fuzzy control", for a survey of fuzzy control cf. [DHR93], e. g. But control is not the only application area of fuzzy logic! This chapter deals with fuzzy applications which do not come under the category fuzzy control. The common framework of the four applications is that only vague observations are available for solving the specific task and that the vagueness is represented by fuzzy sets. But in contrast to fuzzy control no closed loop is to be controlled, i. e. no data flow automatically from the fuzzy system to the "object" under consideration. The "output" of the fuzzy system is shown to a human user or given to a monitoring device, e. g.

In the sequel the four specific tasks being presented in this chapter are sketched:

- In [BB94] a general approach for classification is presented. An expert has defined several classes for a specific domain. The object is characterized by a crisp feature vector, the classification yields a membership vector. Quality inspection, monitoring, technical diagnosis, and medical image processing are mentioned as application examples.
- A chemical compound consisting of different molecules is described in [Ku94] by a model which has four unknown force constants. The constants can not be determined exactly but one knows that they are members of intervals I_1, I_2, I_3, and I_4, resp. Therefore one wants to value each possible combination of constants with a number out of the interval [0,1]. One can measure the constants only indirectly by evaluating the three fundamental vibrations of the oscillating molecule. The fundamental vibrations are a known function **F** of the unknown constants.
- One wants to classify a medical "object", namely an organ, in [KJ94]. An image of the real object is compared with reference images. The object's image as well as the reference images are represented as so-called "wireframe models". Each wireframe model is characterized by a finite set of typical points. The result of the classification is a

membership vector. In order to classifiy the object a spatial transformation of the object's picture is to be performed.

- In [Wi94] a life insurance wants to decide on an application. Possible alternatives are complete acceptance, complete refusal and partial acceptance (only a part of the risks are insured, shorter run-time, e. g.). The decision maker is supported by a collection of n reference applications already decided. The task is to compare the application with the reference applications in order to find out its benefits and drawbacks for the insurance.

2 Classification and Cluster Analysis

All of these tasks can roughly be labeled as classification problems. Related notions are diagnosis (the "object" is a technical device or process or an ill human being, the aim of the classification procedure is to find a suitable repair action or a therapy, and one wants to obtain an explanation for the diagnostic result) and decision making support (one has to make an economical, espec. management, decision and typically has to consider several objective criteria).

We introduce the classification problem formally: Let an object O be given. O is characterized by a t-dimensional feature vector $\underline{x}_O$ of a universe of discourse U. Often U is the space R^t. A set $C_1, \dots, C_n$ of classes is given a priori or has to be discovered. The task is to calculate a membership vector $m_1, \dots, m_n$ for the object O. The number $m_i \in [0,1]$ is the degree of membership of O to C_i. With other words: Each class C_i is characterized by a fuzzy set μ_i of the feature space U, the value $m_i = \mu_i(\underline{x}_O)$ is to be calculated. Ideally $m_i = 1$ for exactly one $i \in \{1,\dots,n\}$ and $m_i = 0$ for all other i holds.

In a lot of practical applications this optimal result can not be derived as one can not observe exactly the features of the object O and/or can not draw a unique conclusion from the features to one class even in the case of a crisp feature vector. If one can only observe a vague feature vector, it is a common approach to represent each observation as a fuzzy set which includes crisp and interval-valued observations.

This approach differs basically from that of classical statistics: Statistics considers random mechanisms. Its outcomes can not be predicted due to random influences but can be observed and measured exactly. Fuzzy sets, however, are a tool to handle imprecise measurements.

Often an expert determines the classes $C_1, \ldots, C_n$ which are to be distinguished. If this is not possible, one can try to obtain the classes with the help of a cluster analysis automatically. N objects $O_1, \ldots, O_N$ are given. The task of a cluster analysis is to find a partition into different classes, called "clusters". Surveys on cluster analysis are [DO74] and [MKB79], e. g. Approaches of the classical cluster analysis require crisp feature vectors and a metric on the feature space. An example for a clustering method is the ISODATA algorithm [DO74]. The distance d(j,k) between the objects O_j and O_k is given. The algorithm iteratively determines a crisp partition into clusters. It finishes if the distance between two clusters always exceeds a given threshold.

Often a crisp partition yielded by a classical method does not allow a practically useful interpretation. In this case one utilizes the more general notion of a fuzzy partition: The degree of membership of object O_j to class C_i is a number $m_i(j) \in [0,1]$. One demands for each object O_j that $m_1(j) + \ldots + m_n(j)$ equals 1 and that each class C_i has at least one "fuzzy member", i. e. there exists an object O_j with a membership degree $m_i(j)$ greater than 0. In order to generate a fuzzy partition for a given set of crisp data one has to define an objective function, a so-called clustering criterion, which values a fuzzy partition. In [Be81], [BH93], and [BN92] fuzzy partitioning algorithms are presented.

If the feature values are no real numbers but fuzzy sets, one additionally needs a metric on fuzzy sets. Fuzzy partitioning algorithms can generalized to fuzzy set valued features also, cf. [BN92].

As mentioned above, the basic classification task is to classify an object O. A set of n classes $\{ C_1, \ldots, C_n \}$ are determined by an expert or automatically by a cluster analysis. The task is to find a membership vector $(m_1, \dots, m_n)$.

The classical analytical approach tries to find explicitely a function **f**: U --> { $C_1, \dots, C_n$ } or more general **f**: U --> $[0,1]^n$ mapping a feature vector to a membership vector. A sample of N classified objects is given. The function **f** is optimized with respect to a suitable criterion. Often the error square sum serves as minimization criterion. Basic solutions of this tasks are the classical regression analysis, polynom classification [Sch77] and neural networks, cf. for example [Ko92] and [KNN93].

The development of knowledge-based systems has stimulated constructive approaches to solve the classification task. A common solution is to formulate IF - THEN - rules (cf. [Su91], e. g.), sometimes denoted by the misleading term "production rules". Such a rule represents a chunk of application-specific classification knowledge. The premise of a rule is a combination of statements on object features. The premise is a statement on the class the object belongs to. Such rules can be weighted with the help of certainty factors as known from the medical diagnostic expert system MYCIN, see [BS84] and [He86]. A usual interpretation of a certainty factor CF is a probabilistic one: CF is a measure of the relative change in the certainty of the rule conclusion which is yielded by knowing that the premise is fulfilled. Different certainty factor calculi are presented in the literature [Su91] but for real-world applications each of them leads to inconsistent, unpredictable, and counterintuitive results [He86].

IF-THEN-rules are the most usual way to organize knowledge for a heuristic classification [Cl85]. An other approach is that of the covering classification [PR90]. One tries to find a class (or a set of classes) which covers the obtaines observations best. Examples for optimization criteria are that of minimization, relevance, and that of highest probability [PR90]. The covering classification is derived in the context of knowledge-based diagnostics.

Other algorithms automatically yield a classification decision tree or a rule set from an example set of classified examples. A survey on algorithms for the construction of classifying decision trees is given in [SL91]. Some approaches utilizes results of information theory. A well-known example is the ID3 classification algorithm [Qu79].

An other modern approach of building a classification tool is to construct a causal network [Su91]. A node represents a proposition, that is a "piece of knowledge" about the object under consideration. An edge points from a cause to an effect and stands for a causal link between two statements. In the presence of uncertain data one often deals with probabilistic causal networks, also called belief networks, where the edges are labeled with weights. In Bayesian belief networks these weights have the interpretation of conditional probabilities. A-priori-probabilities are assigned to the nodes. Belief networks are more flexible than classical causal networks as they can be evaluated in all directions and allow local computations in order to obtain missing data. Bayesian belief networks are treated by [Pe86], [Pe88], [AOJ89], [SS90].

3 Fuzzy Classification and Fuzzy Data Analysis

Several approaches just sketched were generalized to fuzzy set valued data and/or for an imprecise classification.

A method to find explicitely a function **f** : U --> $[0,1]^n$ is described in [BB94]. Each class is represented by a fuzzy set. The function is described as a superposition of these fuzzy sets. The observations are crisp feature vectors.

An other way of solving a classification task is to apply a regression analysis method. A variable **Y** depends on an other variable **X**, it holds y = **f**(x,c) with a known functrion **f** and an unknown parameter c out of a universe of discourse C. In order to estimate the parameter one draws a sample of size N. If one can measure the values of **X** and **Y** only imprecisely, one can consider the sample as a collection of N fuzzy sets $\mu_1, \ldots, \mu_N$ each with the universe of discourse X x Y. The main goal is to calculate a fuzzy set μ_c for the unknown parameter c with C as its universe of discourse. If needed, one defuzzifies this fuzzy set μ_c. According to [BN92] one solves this task in two steps: In the first step the N fuzzy sets are aggregated to one fuzzy set μ_{AGG} on X x Y. The second step is to "extract" μ_c from μ_{AGG}. Suggestions how to perform these steps are given in [BN92], [Ce87], and [Ku90]. A generalized approach considers not only the observations but also the unknown relationship between **X** and **Y** itself as a fuzzy relation. This fuzzy relation is characterized by a membership function on X x Y depending on an unknown parameter c. In [BN92] and [Di88] it is shown how to derive a fuzzy set for c given a sample of N fuzzy sets. Linear regression analysis with a fuzzy sample is treated in [NA90].

In order to illustrate fuzzy data analysis a practical example stemming from [BN92] will be presented in the following. In mining sciences, e. g., one wants to characterize the two-dimensional shape of a three-dimensional particle. The shape is compared with a family of geometric figures, e. g. with generalized ellipses

$$E(c,d,p) = \{ |x/c|^p + |y/d|^p = 1 \mid x \in R \text{ and } y \in R \}.$$

The task is to find a degree of coincidence for each figure, i. e. for each tripel (c,d,p) a value $\mu(c,d,p) \in [0,1]$. One obtains a fuzzy set μ of R_+^3.

The projection of the particle on a plane surface yields a grey-tone picture. The degree of correspondence between this picture and the figure E(c,d,p) is taken as the value $\mu(c,d,p)$. The grey-tone values are mapped to the interval [0,1]. The "fuzzy edge" of the picture is a fuzzy set μ_E on R^2 defined by $\mu_E(x,y) = 2 * \min\{G(x,y), 1-G(x,y)\}$ where G(x,y) is the normalized grey-tone value at the point (x,y). $\mu(c,d,p)$ is calculated as the medium value of μ_E on the curve E(c,d,p) obtained by numerical integration.

The task of classical empirical statistics is to estimate the unknown parameters of a random mechanism with the help of a random sample of size N. This task is also transfered to fuzzy data: If the observations are imprecise, one can consider them as N fuzzy sets being the realizations of a fuzzy set valued random mechanism. In [KM87] and [Vi90]. It is shown how a fuzzy set for the unknown parameters can be calculated by applying statistical methods on a fuzzy set instead of a crisp sample.

Let us turn to some fuzzy versions of the constructive, knowledge-based approach of classification. The well-known concept of a lingustic rule generalizes that of a conventional IF-THEN-rule. The statements in the premise are linguistic instead of crisp statements on an object's features. The conclusion is a crisp or linguistic statement on a membership to a class. The linguistic rule can be weighted by a factor. An example for such a rule is: IF " feature_1 of O is 'big' " AND " feature_2 of O is 'small' THEN " O belongs to class C_2 " with security 0.7.

A fuzzy classificator performs a run-time evaluation of the linguistic rule set as being known from fuzzy control. If a membership vector $(m_1, \ldots, m_n)$ is a sufficient output, a defuzzification is omitted. Such a fuzzy classificator can utilize crisp numbers as well as fuzzy sets as input values.

Two practical applications out of the huge number of fuzzy classificators are sketched. The task in [SG93] is to classify crates; the two classes are "own crate" and "other crate". The classification feature is the printed label of the producer. An other example is given in [We93] and [WPG93]: An automatic transmission adapts automatically to the driver's wishes and his way of driving. The driver as well as the environment conditions are to be classified, and dangerous changings of gears must be prevented. The classificator works as a preprocessor of a controller. Seven input and three output variables (driver class, environment class, dangerous situations) were identified, and the fuzzy classificator was implemented in the car environment.

4 Fuzzy-Neural Classification

Different approaches are suggested in order to combine the advantages of linguistic rules and that of neural networks. Some basic approaches are sketched in the following. The reader is refered to the chapter "some trends in fuzzy control" of this book as well as to [Ko92] and [KNN93].

The first way of combination is just that the neural network serves as a preprocessor for the fuzzy system. Such a preprocessor may be necessary if the feature space is high-dimensional. A stand-alone fuzzy classifier can not be created as the fuzzy system had too much input values or input values without an intuitive semantics. Therefore the neural network calculates the values of intuitive features and reduces and compresses by this the observed data set by abstraction. The output values of the neural network are the input values of the fuzzy system. Based on this approach a hybrid color correction system for HDTVs was constructed [CT93].

In [HFU92] a way is described just to implement a fuzzy controller / classifier of the Sugeno type as a neural network.

The next three approaches utilize a fuzzy-neural combination during the training phase. One way is helpful if the sample does not cover all possible situations. One creates a fuzzy classificator applying linguistic rules. One defines a set of feature vectors not being covered by the sample and let the classificator calculate the corresponding membership vectors. These assignments are added to the sample, and a neural net is adapted to the entire example set.

When creating a fuzzy controller / classifier one has to assign values to a lot of parameters: the shape and parameters of the membership functions, weight factors, sometimes also parameters for the inference method, etc. The idea is that a neural network determines some of these parameters automatically and shortens the engineering phase by this. One way is utilized for example by NeuralGen, a software tool of Togai Infralogic: The user defines the classes, the objects' features, linguistic statements on them, and the membership functions. Each conclusion of a combination of statements to a class forms a possible linguistic rule. The task is to filter out the relevant rules. This is done by a neural network with a suitable architecture calculating the weight factor of each possible linguistic rule.

An other way was suggested by [Ja92] for fuzzy controller / classifier of the Sugeno type and applied in [BZ94], e. g. The parameters of the membership functions (in [Ja92] only Gaussian functions are used) are the only degrees of freedom. They are automatically determined by a suitable neural network called ANFIS (Adaptive Neural Fuzzy Inference System). One estimates roughly these parameters a priori, the neural network adapts them to the given sample.

A fuzzy system interferes the parameters of a neural network during its learning phase in order to accelerate its convergence or to avoid an unwelcome interference of already learned trainig sets by new examples. Typically only few linguistic rules are needed. A simple example for this strategy: The fuzzy system updates the learning rate η of the backpropagation algorithm. This appoach and more complex updating methods are presented in [ACM92] and [HMG94] in the context of fuzzy control. Often these methods turn out to be just "fuzzy formulations" of classical, non-fuzzy procedures. A survey on numerical methods of training acceleration is given in [CU93].

5 Four Fuzzy Classification Applications

In the introduction the four fuzzy application tasks were sketched. In the following it is pointed out what fuzzy sets contribute to the specific solutions and how the authors motivate the utilization of fuzzy sets and fuzzy logic.

The approach of [BB94] is that each class is represented by a fuzzy set with the feature space as its universe of discourse. Each membership function is a so-called "Aizerman's potential function", the classes differ by the parameters of their membership functions. These parameters either stem directly from a human expert or are evaluated with the help of a sample of classified objects utilizing an optimization method. Let $\underline{x}$ be the object's feature vector and μ_i be the membership function of class C_i. The membership degree m_i of the object is just $\mu_i(\underline{x})$.

The method presented in [BB94] competes with methods of the classical statistics, with neural networks, and with expert systems. As an advantage it is stated that an available set of examples can be used as well as the expert's experience. Both sources of knowledge are integrated into the same classificator. The membership functions for the classes illustrate the work of the classificator clearly which is an advantage against weight factors of a neural network. The approach does not require that a specific model is selected or created. The procedure can be applied well even in a high-dimensional feature space.

According to [Ku94] the fundamental vibrations of an oscillating molecule can only be determined approximately. A crisp number pretends a precision which in reality is not available. An interval has two strict borders. Therefore an expert determines a fuzzy set for each fundamental vibration of a molecule with the help of his experience. Only membership functions with a trapezoidal shape are defined. Let μ be the common fuzzy set for the three fundamental vibrations. Let $\mathbf{F} : I_1 \times I_2 \times I_3 \times I_4 \to R^3$ describe the relationship between the force constants and the fundamental vibrations. Then the weight of a combination $\underline{x} \in I_1 \times I_2 \times I_3 \times I_4$ is $\mu[\mathbf{F}(\underline{x})]$. The presented procedure yields a weight factor for each possible combination of force constants being a number out of the interval [0,1]. The result is a fuzzy set with $I_1 \times I_2 \times I_3 \times I_4$ as its universe of discourse which describes a fuzzy relation between the four force constants.

Each characteristic point of a wireframe model presented in [KJ94] is characterized by an own fuzzy set on the space R^3 which takes the value 1 for exactly one tripel. An expert defines three fuzzy sets on the x-, the y-, and the z-axis, resp. Automatically a common fuzzy set on R^3 is calculated. A metric on fuzzy sets is introduced in order to determine the distance between two wireframe models.

A wireframe model with a finite set of characteristic points is only an approximation of an object's picture. Because of this the authors state that it is adequate to represent a point as a fuzzy set.

In [Wi94] the relevant features of the applicant as well as the criteria for evaluating the application are modelled as linguistic variables each having a set of linguistic values as its domain. An expert defines the membership functions and formulates linguistic rules for the inference from the applicant features to the criteria.

In order to compare the application with the reference applications, one needs a value (a number out of [0,1]) for each of the following statements:

- The application p is as good as the application q with respect to criterion j.
- The application p is incompensable worse than the application q with respect to criterion j.

These values are calculated automatically. For this task the fuzzy sets for the linguistic statements are needed as well as two additional fuzzy sets which are defined by an expert quantifying his concept of "as good as" and "incompensable worse", resp. The evaluations are aggregated to a single matrix, called the prevalence matrix, which shows a ranking among the (n+1) applications under consideration.

The criteria which the decision maker must take into consideration typically are fulfilled partly. Therefore they are modelled as linguistic variables. Different criteria can contradict. This conflict of goals can be handled with the help of an outranking procedure. The imprecise knowledge can be integrated smoothly into the procedure when described by fuzzy sets.

An alternative solution is to create an expert system with conventional or linguistic rules or to construct a decision tree. But defining the rule set seems to be time-consuming. In addition, IF-THEN-rules hardly can formalize the pairwise ranking amongst the applications according to different criteria. The task was to support a human decision maker, not to replace him.

6 Conclusion

The four applications show that a good understanding of the underlaying problem is needed. To construct the presented and comparable classificators requires to know the relevant features for the specific applications. One must be able to quantify each feature at least approximately. One must know the classes to be distinguished and must be able to characterize them at least roughly. To acquire and formalize this application-specific knowledge always takes a lot of time. This is demonstrated clearly by the experiences in constructing expert systems. The knowledge acquisition process is known as the "bottle-neck". The experts are always in demand. In addition, the decision making knowledge sometimes must not be disclosed. The available example set often does not cover all situations or has not the structure to be utilized directly.

If the needed knowledge or examples can not be obtained, neither a fuzzy method nor a competing approach will yield a sufficient result. If one is sure to get this knowledge, the next step is to decide what task the classificator has to fulfill. The range of possible answers covers a fully automatical service [BB94] as well as a pure support for a human decision maker [Wi94]. The specific answer determines how reliable and precise the utilized examples and available observations must be and what decisions the tool has to make: for example a ranking amongst the alternatives or a reliable crisp decision.

After having answered this question one has to select an approach out of competing methods, one alternative can be a fuzzy approach. Typically a decision for "fuzzy" does not imply that the complete task is solved by a fuzzy approach - in general a fuzzy solution will perform only a partial task and will cooperate with other methods. Often one has to integrate conventional methods for acquiring and processing measurements and/or interpreting the results. This demonstrates the four applications of this chapter.

Typically one only needs a small part of the large-sized theory of fuzzy sets, fuzzy logic, and approximate reasoning. From my point of view the four articles show this, this observation coincide with that made in [Kr93], e. g. The quality of the results seems not to depend on the precise shape of the membership functions or on the choice of the connecting operators - it is only important to select "sensitive" membership functions and operators. Often one will select alternatives which are easy to handle.

My conclusion is to recommend these steps: If one has a classification, decision making, or control task which has no straightforward solution, one transfers the requirements mentioned above to the own application and check if they are fulfilled. If they do, one compares fuzzy techniques with competing approaches. One has to take into consideration that the validation and the tuning of the solution requires a lot of time. It is always sensitive to define a stopping criterion (budget is spent, time is run-off, a quality criterion is fulfilled). It can save a lot of time to adapt a solution for a similar problem already described in the literature. All the way: the proof of the pudding is in the eating!

Hopefully this book helps its readers in finding an approach which supports them to solve their control, classification, or decision making task!

References

[ACM92] P. ARABSHAHI, J. J. CHOI, J. MARKS & T. P. CANDELL: "Fuzzy Control of Backpropagation". Proceed. IEEE Intern. Conf. on Fuzzy Systems, San Diego, 1992.

[AOJ89] S. K. ANDERSEN, K. G. OLESEN, F. V. JENSEN & F. JENSEN: "HUGIN - a Shell for Building Bayesian Belief Universes for Expert Systems". Proceed. 11th Int. Conf. Artif. Intell., Detroit, 1989.

[Be81] J. C. BEZDEK: "Pattern Recognition with Fuzzy Objective Function Algorithm". Plenum Press, New York, 1981.

[BB94] S. BOCKLISCH & N. BITTERLICH: "Fuzzy Pattern Classification - Methodology and Applications". In: R. KRUSE, J. GEBHARDT & R. PALM (eds.): "Fuzzy Systems in

Computer Science". Vieweg Verlag, Braunschweig, 1994.

[BH93] J. C. BEZDEK & J. D. HARRIS: "Convex Decompositions of Fuzzy Partitions". In: D. DUBOIS, H. PRADE, R. R. YAGER (eds.): "Fuzzy Sets for Intelligent Systems". Morgan Kaufmann Publ., San Mateo, 1993, pp. 617 - 628.

[BN92] H. BANDEMER & W. NÄTHER: "Fuzzy Data Analysis". Kluwer Publ., Dordrecht, 1992.

[BS84] B. G. BUCHANAN & E. SHORTLIFFE (eds.): "Rule-Based Expert Systems: The MYCIN Experiments on the Stanford Heuristic Programming Project". Addison Wesley, New York, 1984.

[BZ94] K. BRAHIM & A. ZELL: "ANFIS-SNSS: Adaptive Network Fuzzy Inference System in the 'Stuttgarter Neuronale Netze Simulator' ". In: R. KRUSE, J. GEBHARDT & R. PALM (eds.): "Fuzzy Systems in Computer Science". Vieweg Verlag, Braunschweig, 1994

[Ce87] A. CELMINS: "Least Squares Model Fitting to Fuzzy Vector Data". Fuzzy Sets and Systems 22 (1987), pp. 245 - 269.

[Cl85] W. CLANCEY: "Heuristic Classification". Artif. Intell. 27 (1985), pp. 289 - 350.

[CT93] P.-R. CHANG & C. C. TAI: "Model-Reference Neural Color Correction for HDTV-Systems based on Fuzzy Information Criteria". Proceed. Intern. Conf. on Fuzzy Systems, San Francisco, 1993.

[CU93] A. CICHOCKI & R. UNBEHAUEN: "Neural Networks for Optimisation and Signal Processing". Teubner & Wiley, Stuttgart & Chichester, 1993.

[Di88] Ph. DIAMOND: "Fuzzy Least Squares". Inform. Sc. 46 (1988), pp. 141 - 157.

[DHR93] D. DRIANKOV, H. HELLENDOORN & M. REINFRANK: "An Introduction to Fuzzy Control". Springer Verlag, Heidelberg, 1993.

[DO74] S. B. DURAN & P. L. ODELL: "Cluster Analysis: A Survey". Springer Verlag, Berlin, 1974.

[DPY93] D. DUBOIS, H. PRADE & R. R. YAGER (eds.): "Fuzzy Sets for Intelligent Systems". Morgan Kaufmann, San Mateo, 1993.

[He86]: D. E. HECKERMAN: "Probabilistic Interpretations for MYCIN's Certainty Factors". In: L. KANAL & J. LEMMER (eds.): "Uncertainty in Artificial Intelligence". North Holland Publ., Amsterdam, 1986, pp. 167 - 196.

[HFU92] S. HORIKAWA, T. FURUHASHI & Y. USHIKAWA: "On Fuzzy Modeling using Fuzzy Neural Networks with the Backpropagation Algorithm". IEEE Transact. on Neural Networks, Vol. 3, No. 5 (1992).

[HMG94] S. K. HALGAMUGE, A. MARI & M GLESNER: "Fast Perceptron Learning by Fuzzy Controlled Dynamic Adaption of Network Parameters". In: R. KRUSE, J. GEBHARDT & R. PALM (eds.): "Fuzzy Systems in Computer Science". Vieweg Verlag, Braunschweig, 1994.

[Ja92] J.-S. ROGER JANG: "Self-Learning Fuzzy Controllers based on Temporal Backpropagation". IEEE Transact. on Neural Networks, Vol. 3, No. 5 (1992), pp. 714 - 723.

[Ko88]: Y. KODRATOFF: "Introduction to Machine Learning". Pitman, 1988.

[Ko92] B. KOSKO: "Neural Networks and Fuzzy Systems". Prentice Hall, Englewood Cliffs, 1992.

[Kr93] R. KRUSE: "Fuzzy sets - einige Klarstellungen". KI 4/93 (1993), S. 73 - 74.

[Ku90] M. KUDRA: "Comparison of Methods to Evaluate Fuzzy Parameters in Explicit Functional Relationships". Freiberger Forschungshefte D 197 (1990), Dt. Verlag f. Grundstoffindustrie, pp. 43 - 58.

[Ku94] M. KUDRA: "A Fuzzy Method to Spectra Interpretation". In: R. KRUSE, J. GEBHARDT & R. PALM (eds.): "Fuzzy Systems in Computer Science". Vieweg Verlag, Braunschweig, 1994.

[KB93] D. H. KRAFT & D. A. BUELL: "Fuzzy Sets and Generalized Boolean Retrieval Systems". In: In: D. DUBOIS, H. PRADE, R. R. YAGER (eds.): "Fuzzy Sets for Intelligent Systems". Morgan Kaufmann, San Mateo, 1993, pp. 648 - 659.

[KJ94] L. KÖHLER & P. JENSCH: "Fuzzy Elastic Matching of Medical Objects Using Fuzzy Geometric Representations". In: R. KRUSE, J. GEBHARDT & R. PALM (eds.): "Fuzzy Systems in Computer Science". Vieweg Verlag, Braunschweig, 1994.

[KKN93] F. KLAWONN, R. KRUSE & D. NAUCK: "Neuronale Netze in der KI". Vieweg Verlag, Braunschweig, 1993.

[KM87] R. KRUSE & K. D. MEYER: "Statistics With Vague Data". Reidel, Dordrecht, 1987.

[MKB79] K. V. MARDIA, J. T. KENT & J. M. BIBBY: "Multivariate Analysis". Academic Press, London, 1979.

[NA90] W. NÄTHER & M. ALBRECHT: "Linear Regression with Random Fuzzy Observations". Statistics 21 (1990), pp. 521 - 531.

[Pe86] J. PEARL: "Fusion, Propagation, and Structuring in Belief Networks". Artific. Intell., Vol. 29 (1986), pp. 241 - 288.

[Pe88] J. PEARL: "Probabilistic Reasoning in Intelligent Systems: Networks of Plausible Inference". Morgan Kaufmann Publ., San Mateo, 1988.

[PR90] YUN PENG & J. REGGIA: "Abductive Inference Models for Diagnostic Problem-Solving". Springer Series Symbolic Computation - Artificial Intelligence". Springer Verlag, Heidelberg, 1990.

[Qu79] J. R. QUINLAN: "Discovering Rules from Large Collections of Examples: A Case Study". In: D. MICHIE (ed.): "Expert Systems in the Micro-Electronic Age". Edinburgh University Press, 1979.

[Ru93] E. H. RUSPINI: "Numerical Methods für Fuzzy Clustering". In: D. DUBOIS, H. PRADE, R. R. YAGER (eds.): "Fuzzy Sets for Intelligent Systems". Morgan Kaufmann Publ., San Mateo, 1993, pp. 599 - 614.

[Sch77] J. SCHÜRMANN: "Polynomklassifikatoren für die Zeichenerkennung". Oldenbourg Verlag, München, 1977.

[SG93] D. SCHODER & H. GEIGER: "Das Geheimnis um Neuro-Fuzzy - Künstliche Intelligenzen profitieren voneinander". Elektronik plus 2/93 (1993), S. 123 - 127.

[SL91] S. R. SAFAVIAN & D. LANDGREBE: "A Survey of Decision Tree Classifier Methodology". IEEE Transact. Systems, Man & Cybern., Vol. 21, No. 3 (1991), pp. 660 - 674.

[SS90] P. SHENOY, G. SHAFER: "Axioms for Probability and Belief-Function Propagation". In: R. D. SHACHTER, T. S. LEVITT, L. N. KANAL & J. F. LEMMER (eds.): "Uncertainty in Artificial Intelligence", Vol. 4. North Holland Publ., Amsterdam, 1990.

[Su91] K. SUNDERMEYER: "Knowledge-Based Systems - Terminology and References". BI Wissenschaftsverlag, Mannheim, 1991.

[Vi90] R. VIERTL: "Einführung in die Statistik". Kap. VII: "Statistische Analyse für unscharfe Daten". Springer Verlag, Wien, 1990.

[We93] H.-G. WEIL: "Einfühlsames Automobil - Schaltlogik mit integriertem Fuzzy-Regler für ein Automatikgetriebe". Elektronik plus 2/93 (1993), S. 91 - 94.

[WPG93] H.-G. WEIL, G. PROBST & F. GRAF: "Fuzzy Shift Logic for an Automatic Transmission System". Proceed. EUFIT '93, Aachen, 1993.

[Wi94] M. WIEMERS: "FRED (Fuzzy Preference Decision Support System) - a Knowledge-Based Approach for Fuzzy Multiattribute Preference Decision Making". In: R. KRUSE, J. GEBHARDT & R. PALM (eds.): "Fuzzy Systems in Computer Science". Vieweg Verlag, Braunschweig, 1994.

6.2. Fuzzy Pattern Classification – Methodology and Application –

Steffen F. Bocklisch and Norman Bitterlich

1 Introduction

It is always impressive to watch an experienced specialist master difficult and complex situations. The real problems are very different but the characteristics of the action are similar: acquiring a flood of information, concentrating on the essential, evaluating and reaching a purpuseful decision. The expert often cant explain his decision making process. Doubtlessly, his knowledge and experiences about the functional connections are very important. However his ability to compare the present situation with typical examples leads to a strategy in a more effective way. This intellectual process means an immense reduction from a lot of information to a few pattern.

The principles of classification can be used to automate this thinking manner. Corresponding to the real experiences, the vagueness and uncertainty are awarely noticed and used for adequate modeling. Methods based on fuzzy set theory assume suitably for describing and processing inexact information and vague knowledge in a mathematical sense. Problems in decision making and control which are difficult or impossible to access by theoretical analysis appear solvable. Knowledge and experiment based methods complete the process modeling successfully.

2 Methodology of Fuzzy Pattern Classification

The research in the field of fuzzy technologies in Chemnitz already started in the early seventies [Peschel, Bocklisch 1977]. The principles of classification had been combined with the main ideas of fuzzy set theory. The "Fuzzy Pattern Classification" based on a pattern concept was created [Bocklisch 1987] and has opened a qualitative new way of process modeling.

The classification model is represented by a set of classes made of fuzzy subsets of an information space. This space can be high-dimensional and is defined by features, i. e., by process information which can probably lead to the classification. Such features can be direct measurements or derived values resulting from signal processing. In this case vagueness appears as measuring fault or process disturbance. In a similar way, indirect measuring causes further uncertainties. Moreover, verbal description with all its subject and linguistic indefiniteness can be regarded as features.

All defined classes have their own meaning. So quality, fault or control classes are likely. The expert defines they depend on the goal of classification. A class consists of similar objects identified with its corresponding feature vectors of parameters. Despite these parameters, the global rate of process situation can be inserted into the structure building. For example, already known process states, acoustical and visual perceptions or the actual process conditions and settings include a lot of additional information. The expert can interpret the classes, so he can connect them with certain actions, i. e. quality ratings or control strategies.

According to the fuzzy set theory, each class is defined by a subset of membership functions within the information space. A concept of parametric functions was uniformly chosen taking the general Aizerman potential function as basis. In the one-dimensional case, this function has the following form:

$$\mu(x) = \frac{1}{1+(\frac{1}{b}-1)\cdot|\frac{x-s}{c}|^d} \text{ für } x \in R$$

The function is unimodal and can be adapted to the given actual problem by tuning the parameters s, c, b and d, which can be defined as:

s ... the gravity point of the class for degree of membership 1,
c ... the boundary of the class,
b ... the membership on the class boundary, i.e. $\mu(s+c) = \mu(s-c) = b$
d ... the shape parameter for decrease in membership with increasing distance from s.

To increase the flexibility of adaptation this representation can be generalized by distinguishing a right and a left wing of the function connected at s. In this way, even unsymmetrical parameters (different for the right and left wing) can be taken into account. The aggregation operation for the one-dimensional membership function is basically a disjunction one. The n-dimensional relation follows the parametrical concept:

$$\mu(x_1, \ldots, x_n) = \frac{1}{1+\sum_{i=1}^{n} (\frac{1}{b_i}-1) \cdot |\frac{x_i - s_i}{c_i}|^{d_i}} \text{ für } (x_1, \ldots, x_n) \in R^n$$

In this form, the term is very simple. It does not represent the proper vector processing. However, this aspect enables regarding and modeling the dependencies between several features. Only this possibility ensures the increased efficiency of this classification model.

3 The classifier design

Each object is described by a feature vector. To conclude from a set of measurements (training data) to a global class description, a learning period is necessary. Generally the classifier design contains the following steps:

An exhaustive process analysis in close cooperation with specialists has to be at the beginning. Hence the problem and the classification object must be specified. The goal determines the classes and the features with the appropriate measured values. The selection can be supported by analytical process knowledge. Also, the imitation of human actions can be useful to the problem solving. Repeatedly, the practical experiences and the theoretical information complete each other advantageously. Subsequent to these decisions, the technical conditions for measuring have to be realized and the active or passive experiments have to be planed.

The choice of the features influences the efficiency of the classification model considerably. Conventional methods or particular algorithms of data signal processing find broad use. During the experiments, the training data will be recorded and the feature vector will be computed.

The collected learning objects are assigned to the predefined classes. In completion of the "learning by teaching" procedure, mathematical clustering techniques can be used.The membership function of the various classes based on the parameters of the associate learning objects will be computed. For this purpose the feature data of objects which belong together can be processed by the comfortable software "FUCS - Fuzzy Classification System" [Bocklisch et al. 1992a].

The learning phase is completed with tests and simulations to verify the efficiency of the classification model. For that reason the reclassification rate is regarded as an important criterion. It specifies the percentage of correct classified learning objects. However, other quality criteria are possible to characterize certain performances of the classification model.

4 The classifier optimization

During the learning phase, the classifier design can be influenced in various manners. The selected classes and the determined structure are of crucial importance. Due to the well-known parameters of the membership functions, subjective manipulation can modify the classification model purposefully now. Such interventions in the design can be necessary in the case that some classes are not sufficiently represented by learning objects or if suitable experiments are not applicable. Especially for modeling classes, such as disturbances, technological limits or alarm situations, it seems to be necessary and possible to pay attention to rule-based knowledge.

Which features have to be used should be evaluated carefully. Surely they contribute to the classification differently. That's why it is interesting to find out the significant features. The objective justified reduction of feature numbers [Priber 1989] cuts the expense of an classification process. So the computing time decreases directly. However real costs can also be cut down by renouncing some sensors and measuring techniques.

Finally the best suitable classification model is chosen from several alternative designs.

5 The usage of the classifier

The whole knowledge derived from experiences and experiments is compressed into parameters of the fuzzy classifier. Now it is used to classify an unknown object represented by its feature vector. All classes respond to a certain extent. So to speak, they "fire" corresponding to the computed degree of membership to each class. The exact results of the degree of membership of all classes are combined in the so-called sympathy vector. So, the identification leads from the high-dimensional feature vector to the low-dimensional and interpretable sympathy vector.

The fuzzy consideration does not only bring out the degree of membership to classes but also the risk of this decision:
- If the degree of membership is obviously higher to one of the classes than to other ones, the decision is quite risk-free.
- The membership to a class is uncertain when two (or more) degrees are high without quantitative differences.
- If the degree of membership to none of the classes is high enough or does not exceed a fixed identification boundary, the membership to one of the classes is not likely. The object is turned out.

6 Applications

The methodology of fuzzy pattern classification is very general. Thus, it is applicable in a wide range of applications, like process identification, modeling or decision making. In 1985 the efficiency of the procedure was already proved successful in practice.
- In technical field: Monitoring of wear fault of conveyor belt rolls for the predicted maintenance [Bocklisch et al. 1989].
- In medical field: Interpretation of radiographs for the diagnosis with respect to early stages of carcinoma of the mammary gland [Schüler 1993].

These applications exhibit a certain similarity: Before using fuzzy classification based methods extensive experiences are necessary to compare the noise or pictures by subjective acoustical or visual perception respectively.

In the field of quality inspection and assurance, numerous projects verified the procedure, too. The following list remains, however, incompletely:
- Based on vibroacoustical analysis, the tool sharpness (drilling and milling cutter) was determined on-line in flexible production. Because of

technological measuring restrictions, only indirect diagnosis signals were available [Bocklisch et. al. 1992b].
- The wear fault of rolling disks was observed to early detect cracks by monitoring vibration signals [Hänel, Luft 1991].
- The robot MAG welding quality of robot welding technique was assured by analyzing electrical current and voltage curves [Burmeister 1992].
- Defining the quality of refrigerators by investigating the time series during the start phases of the cooling process [Bocklisch, Lorenz 1990].
- Noise estimation for quality defining [Bitterlich, Totzauer 1992].
- Quality inspection of polished surfaces with methods of image processing [Priber 1991].

Other fields of application are:
- the exploitative data analysis,
- the pattern recognition in general image domain (analysis of images, time series or frequencies),
- the technical and non-technical diagnosis, or
- a class-based control.

Though, the design strategy and the application conditions are generally applicable, the attention can be directed to different parts. Some examples explain this fact: The data analysis requires an intelligent data preprocessing. The numbers of data increase in multi-sensor systems. Time series must be analyzed in real time. The system complexity in technical or ecological monitoring increases. That all requires new performance of both software and hardware [Bocklisch, Müller et. al. 1992].

References

[Bitterlich, Totzauer 1992] Bitterlich, N.; Totzauer, R.: Geräuschbewertung mittels Fuzzy Klassifikation. Proc. 6. Kongreßmesse für industrielle Meßtechnik "MessComp'92", Wiesbaden, 1992, S. 140-140

[Bocklisch 1987] Bocklisch, S.F.: Prozeßanalyse mit unscharfen Verfahren. Verlag Technik, Berlin 1987

[Bocklisch et al. 1989] Bocklisch, S.F.; Fuchs, T.; Meltzer, G.: Objektivierte industrielle Zustandsdiagnostik von rotierenden Baugruppen, erprobt an Tragrollenstationen von Gurtbandförderern. J. Maschinenbautechnik 38 (1989), 2, S. 79-82

[Bocklisch et al. 1992a] Bocklisch, S.F.; Burmeister, J.; Priber, U.: FUCS: Diagnose, Überwachung und Steuerung von Prozessen mit Hilfe der unscharfen Klassifikation rechnergestützt. J. Der Maschinenschaden 65 (1992), 1, S. 26-29

[Bocklisch et al. 1992b] Bocklisch, S.F.; Bitterlich, N.; Ruhnau, G.: On-line-Überwachungssystem - eingesetzt zur Verschleißdiagnostik an Drehmaschinen. Proc. Fachtagung "Automatisierung", Dresden, Vol. 5, S. 125-130

[Bocklisch, Lorenz 1990] Bocklisch, S.F.; Lorenz, A.: Diagnosegestützte Zuverlässigkeitsabschätzung von Haushaltkühlschränken mittels der Methode der unscharfen Klassifikation. Proc. Symposium "Zuverlässigkeit", Magdeburg, 1990

[Bocklisch, Müller et al. 1992] Bocklisch, S.F.; Müller, D. et al.: Ein Akzeleratorsystem zur unscharfen Klassifikation - Theorie und Realisierung. Proc. VDI-Fachtagung "Technische Anwendung von Fuzzy-Systemen", Dortmund, 1992

[Burmeister 1992] Burmeister, J.: Fuzzy Logic - nicht nur ein Modetrend. J. Schweißen und Schneiden 43 (1992), 9, S. 533-537

[Hänel, Luft 1991] Hänel, U.; Luft, M.: Ermittlung von Diagnoseparametern zur Erkennung von Walzscheibenanrissen mit Hilfe des Programms FUCS. J. Neue Hütte (1991), 2, S. 57-62

[Peschel, Bocklisch 1977] Peschel, M.; Bocklisch, S.F.: Unscharfe Modellbildung und Steuerung. Fortschrittsberichte GDR-Working Group for Fuzzy Sets and Systems, Vol. I - VI, TH Karl-Marx-Stadt (Chemnitz)

[Priber 1989] Priber, U.: Ein Verfahren zur Merkmalsreduktion für unscharfe Klassifikatoren. Diss. A, TU Karl-Marx-Stadt (Chemnitz), 1989

[Priber 1991] Priber, U.: Oberflächenprüfung mittels Beugungslichtsensor und unscharfe Klassifikation. Proc. 7. Tagung "Meßinformationssysteme", Chemnitz, 1991

[Schüler 1993] Schüler, W.: Specification of Medical Data With Fuzzy Methods. In: Bandemer, H. (eds.):Modelling Uncertain Data. Akademie Verlag, Berlin, 1993

6.3. A Fuzzy Method to Spectra Interpretation

Matthias Kudra

Abstract

The purpose of this paper is to introduce another application of fuzzy set theory on the field of spectra simulation. Spectra are characterized by a lot of bands. Simulation means to forecast the frequency positions of the bands. To do this, the geometry of the molecule and so-called force constants, reflecting intermoleculare bonds, have to be known. However, the numerical values of the force constants are known only aprroximately and partially in most cases. Since the application of iteration methods like least square methods to determine moleculare force fields is bounded, we will propose an alternative way. Starting from a fuzzy interpretation of uncertainty, we develop a completation of the traditional method to the calculation of vibrational frequencies: the Normal Coordinate Analysis (NCA) The method will be demonstrated and discussed on the example of 3- atomic molecules like water H_2O, hydrogensulfid H_2S and hydrogen-selenium H_2Se.

1 Vibrational Spectra and Normal Coordinate Analysis

In chemistry, the spectroscopy is an very important method to investigate combine and bond behaviour of moleculare systems. The spectral bands, resulted by infrared or RAMAN measurements, mark such frequencies of the light absorbed by the molecule to the activation of vibrations. They are characterized by two parameters: band position and intensity. With the help of calculation methods we are able to assign the vibrations of atomic groups to special bands in the spectra. The Normal Coordinate Analysis (NCA) is such an important method to calculate the band positions of so- called fundamental or normal vibrations. As a rule the fundamental vibrations have the highest intensity in the spectra. The NCA starts from a mechanical point of view and model the atoms as point masses and the bonds by assuming of spring forces.

As a result we get the eigenfrequencies and eigenvectors of the fundamental vibrations. With the help of the eigenvectors we can assign the eigenfrequencies to the experimental measured frequencies of the fundamental vibrations. To the calculation of the eigenfrequencies we need geometry parameters (bond distances and angles), the atomic masses and especially so-called force constants reflecting the strength of the various bonds and interactions between them. An overview about NCA is given by [McIntosh, Michaelian 1979], computer programs are developed by [Jones 1976].

[Nibler, Pimentel 1968] discuss the special problem of the 3- atomic molecules H_2O, H_2S and H_2Se. The geometry of these molecules is given by the atomic distance r: H-O, H-S, H-Se, resp., and the angle d between the atoms. The interactions between the atoms are described by 4 force constants: Fr, Frr, Frd and Fd. The force constants Frr and Frd stand for the couplings between Fr and Fd, describing stretches and angle bending, resp.. Hence, we get the following 3 normal vibrations illustrated in figure 1: symmetric (1) and asymmetric (2) stretch and angle bending (3).

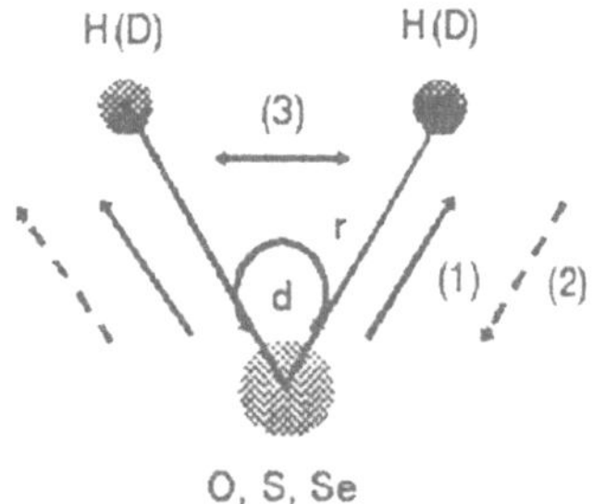

For H_2O it holds:

r = 0.958 Å

d = 104.45°

Figure 1: Normal vibrations of the 3-atomic molecules

However, in many cases we have only vage informations about the numerical values of the NCA input parameters. Especially, this comes true for the force constants. Frequently, they are known only approximately and partially. Even in the case of the 3- atomic molecules the force field it is not known without any doubt. The geometry parameters are approximate values too, because in reality no molecule is in steady position. However, the uncertainty of the geometry can be neglected according to the uncertainty of the force constants, because from special measurements (X-ray investigations) the geometry parameters are known with sufficent precision.

To get an optimal aggreement between the model (NCA) and the reality (spectra), frequently an iterative improvement of the force field is necessary. Additionally, the iteration needs the observed experimental frequencies of the fundamental vibrations. However, the experimental values are uncertain quantities, too. Sometimes we can only determine frequency intervals expecting the normal vibrations. Furthermore, because of the overlapping of vibrations and the appearence of combination vibrations, we cannot give an unique assignment of experimental frequencies to normal vibrations. Especially this comes true for complex moleculare structures like zeolites. Furthermore, to carry out an iterative improvement of force constants by classical least sqares methods (LSE), all experimental frequencies have to known exactly. The LSE use the euclidic distance measure between calculated and observed frequencies as an evaluation measure of force fields. Furthermore, the LSE assume that the number of force constants is smaller than the number of experimental frequencies. However, this assumption is failed for 3- atomic molecules, where we have 4 force constants and 3 observed experimental frequencies. Hence, to circumvent this fact, we can fix some force constants or we can include additionally quantities like experimental frequencies of isotope molecules. Mentioned that for isotope molecules some H- atoms are substituted by D- atoms. In the light of NCA, the only distinguish between the origin molecules and its isotopes lies in the mass of the D- atom, because in following of the Born- Oppenheiner- Theory the same force field can be used for all isotopes. However, since the experimental frequencies of isotope molecules are more uncertain, the isotope method has its boundaries, too.

Hence, we will propose an alternative method removing LSE drawbacks. We start from a quite different point of view and model theoretical as well as experimental uncertainty by fuzzy sets and subtitute the euclidic distance measure by a fuzzy measure.

2 Fuzzy observations of the experimental frequencies and force constant intervals

In the case of the 3- atomic molecules we are able to fix the band positions of the fundamental vibrations. However, the spectral bands are not determined exactly. There are some uncertainty regions, where the fundamental vibrations can also take their frequencies. Furthermore, from theoretical point of view it is sufficient to calculate the eigenfrequencies so that they have at most a 3% deviation from the experimental ones. Hence, about the fundamental vibrations of the water molecule we can use the following experimental information.

TABLE 1
Experimental observed frequencies and uncertainty regions of the normal vibrations in cm^{-1}. (condense phase)

Isotope	asymmetric	symmetric	angle
	O-H stretch		bending
H_2O	3755.8	3656.7	1594.6
	[3634, 3868]	[3547, 3766]	[1547, 1642]
D_2O	2788.1	2671.5	1178.3
	[2704, 2872]	[2591, 2752]	[1143, 1214]
HDO	3707.5	2726.7	1403.4
	[3596, 3819]	[2645, 2809]	[1361, 1446]

The observed experimental frequencies are the centers of the uncertainty regions expressed by intervals. In the uncertainty regions we will prefer frequencies *near by* the interval centers against frequencies at the boundaries. This is the starting point of fuzzification. In the simpelst case, we can evaluate the observed experimental frequencies with 1, the values on the interval boundaries with 0 and connect the points by a straight line. The arising fuzzy sets are triangular fuzzy numbers. There membership functions are given by:

$$m_{\tilde{v}_{ij}}(f) = \begin{cases} \dfrac{(f - l_{ij})}{\left|e_{ij} - r_{ij}\right|}, & \text{if } l_{ij} \le f \le r_{ij} \\[2ex] \dfrac{(r_{ij} - f)}{\left|e_{ij} - l_{ij}\right|}, & \text{if } e_{ij} \le f \le r_{ij} \\[2ex] 0, & \text{otherwise} \end{cases} \tag{1}$$

The index i stands for the isotope number: 1 for H_2O, 2 for D_2O and 3 for HDO and the index j describes the type of the experimental frequence (1 and 2 for asymmetric and symmetric O- H stretches, respectively, and 3 for angle bending). The quantities e_{ij} are the experimental frequencies in the i-th row bending). The quantities e_{ij} are the experimental frequencies in the i-th row and j-th column of table 1 and r_{ij}, l_{ij} are the corresponding interval boundaries. We will call the fuzzy numbers *fuzzy observations of the experimental frequencies*. The membership functions of the fuzzy observations are illustrated in figure 2 for the H_2O- molecule..

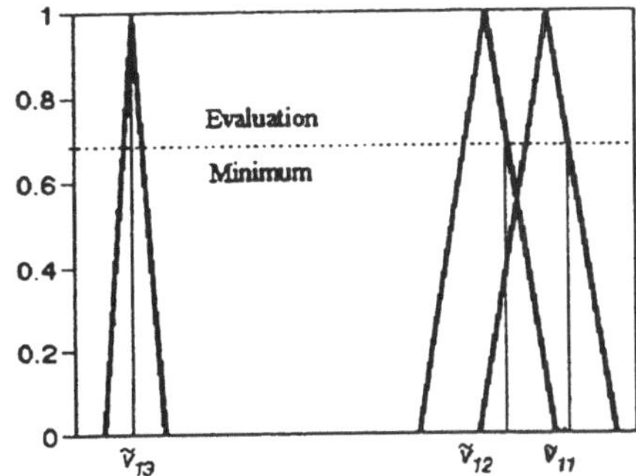

Figure 2: Fuzzy Observation and Evaluation for H_2O

Summarized all experimental information about band positions and uncertainty of the fundamental vibrations is expressed by fuzzy observations. Now, we can collect the single fuzzy observations $\tilde{\nu}_{i1}, \tilde{\nu}_{i2}$ and $\tilde{\nu}_{i3}$ for the i-th isotope to one fuzzy observation $\tilde{\nu}_i$ containing the same information as the single fuzzy sets. This is achieved by cross product of the fuzzy sets: $\tilde{\nu}_i = \tilde{\nu}_{i1} \times \tilde{\nu}_{i2} \times \tilde{\nu}_{i3}$.

If the cross product is defined by the minimum operator, the membership functions of the fuzzy sets $\tilde{\nu}_i$, i=1(1)3, are given by:

$$m_{\tilde{\nu}_i}(\nu_{i1}, \nu_{i2}, \nu_{i3}) = \min(m_{\tilde{\nu}_{i1}}(\nu_{i1}), m_{\tilde{\nu}_{i2}}(\nu_{i2}), m_{\tilde{\nu}_{i3}}(\nu_{i3})) \tag{2}$$

Equation (2) assigns to every vector $(\nu_{i1}, \nu_{i2}, \nu_{i3})$ of experimental values a fuzzy evaluation. The assignment is illustrated in figure 2 by a doted line. On the same way, we combine the single fuzzy observations for each isotope to one fuzzy set $\tilde{\nu} = \tilde{\nu}_1 \times \tilde{\nu}_2 \times \tilde{\nu}_3$. Starting from (2) the membership function is given by:

$$m_{\tilde{\nu}}(\nu_1, \nu_2, \nu_3) = \min(m_{\tilde{\nu}_1}(\nu_1), m_{\tilde{\nu}_2}(\nu_2), m_{\tilde{\nu}_3}(\nu_3)) \tag{3}$$

From a heuristical point of view, it is all the same, if the single fuzzy observations or the combined ones are used.

From physical point of view, unprecise informations about the force constants are available in form of interval restrictions. We can determine special intervals where the force constants take their values. Values outside of the intervals can be neglected. Contrary to the experimental frequencies, we cannot prefer some values inside of the intervals. Hence, at first we will take ordinary (crisp) intervals to model the uncertainty of force constants. However, we will see at the end of this paper that sometimes it is useful to model the uncertainty of force constants by fuzzy numbers, too.

For example, the uncertainty of the 4 water force constants can be expressed by the following interval restrictions:

$$\begin{array}{ll} Fr \in I_r = [6.5, 8.5] \text{ mdyn/ Å} & Frr \in I_{rr} = [-1.5, 1.0] \text{ mdyn/ Å} \\ Fd \in I_d = [0.6, 2.0] \text{ mdyn Å} & Frd \in I_{rd} = [-1.5, 2.0] \text{ mdyn.} \end{array} \tag{4}$$

Now, the consideration of interval uncertainty means to carry out NCA for all force constant combinations $(Fr, Frr, Frd, Fd) \in I_r \times I_{rr} \times I_{rd} \times I_d$.

Since the number of possible force constant combinations is not enumerable, we calculate with all combinations resulting by a discretization of the intervals. For example, we can take an equal step width of 0.05 in the intervals I_r, I_{rr}, I_{rd} and I_d. Then, 4,305369 million combinations arises. The immense number of combinations shows that a solution of such problems require powerful fast computers.

3 Fuzzy Evaluation

Now, we can define the fuzzy evaluation of a force constant combination. This is the significant step of fuzzification. To the evaluation we start from the calculated eigenfrequencies and consider their membership degrees to the corresponded fuzzy observations of the experimental frequencies. According to (2) and (3) we have to substitute the quantities v_{ij} by the corresponded calculuated eigenfrequencies. First of all we have to assign the quantities v_{ij} to the their corresponded eigenfrequencies by means of eigenvectors. In the case of the 3- atomic molecules, the sign of the first two eigenvector components must be the same for symmetric stretch and must be opposite for asymmetric stretch. In general, we assign the quantities v_{i1} and v_{i2} to asymmetric and symmetric stretches, respectively, and v_{i3} to angle bending. However, there are some cases where v_{i1} and v_{i2} change there roles.

The fuzzy evaluation has to carry out for all force constant combinations resulting by the interval discretizations. Combinations with a high fuzzy evaluation (their membership degree is near by 1) have a good frequency approximation in the fuzzy sense. According (2) and (3), the following fuzzy evaluations can be distinguished:

- the *single fuzzy observation*, if only the experimenatl frequencies of one isotope are taken into consideration
- the *total fuzzy evaluation*, if the experimental frequencies of all isotopes are taken into consideration.

Additionally, we will introduce an *isotope evaluation* as a measure to evaluate the similarity of single fuzzy evaluations. The isotope evaluation is of high physical importance, because all isotopes of an given molecule are descibed by the same force field. We have to emphasize that to a fuzzy evaluation of force fields the consideration of isotope molecules is optional.

The single fuzzy evaluation for the H_2O- molecule called B_1 is defined by:

$$B_1(\mathrm{Fr}, \mathrm{Frr}, \mathrm{Frd}, \mathrm{Fd}) = \min(m_{\tilde{\nu}_{11}}(\nu_{11}), m_{\tilde{\nu}_{12}}(\nu_{12}), m_{\tilde{\nu}_{13}}(\nu_{13}))$$

where ν_{11}, ν_{12} and ν_{13} are the calculated eigenfrequencies of the force field (Fr, Frr, Frd, Fd). Analogously, it follows the single fuzzy evaluations B_2 for D_2O and B_3 for HDO. The total fuzzy evaluation called B is defined by

$$B(\mathrm{Fr},\mathrm{Frr},\mathrm{Frd},\mathrm{Fd}) = m_{\tilde{\nu}}(\nu_1, \nu_2, \nu_3) = \min(m_{\tilde{\nu}_1}(\nu_1), m_{\tilde{\nu}_2}(\nu_2), m_{\tilde{\nu}_3}(\nu_3)) = \min(B_1, B_2, B_3)$$

The isotope evaluation as a measure like the fuzzy sets B_1, B_2 and B_3 are similar can be defined by:

$$B_I = \frac{\min(B_1, B_2 B_3)}{\max(B_1, B_2 B_3)} = \frac{B}{\max(B_1, B_2 B_3)}$$

The fuzzy evaluations B_1, B_2, B_3, B and B_I can be interpreted as *fuzzy relations* on the force constant space, stretched by the force constant intervals. The projections of the relations to the axes Fr, Frr, Frd and Fd are a main tool to analyze the relations. In the case of the water molecule we get the projections illustrated in figure 3.

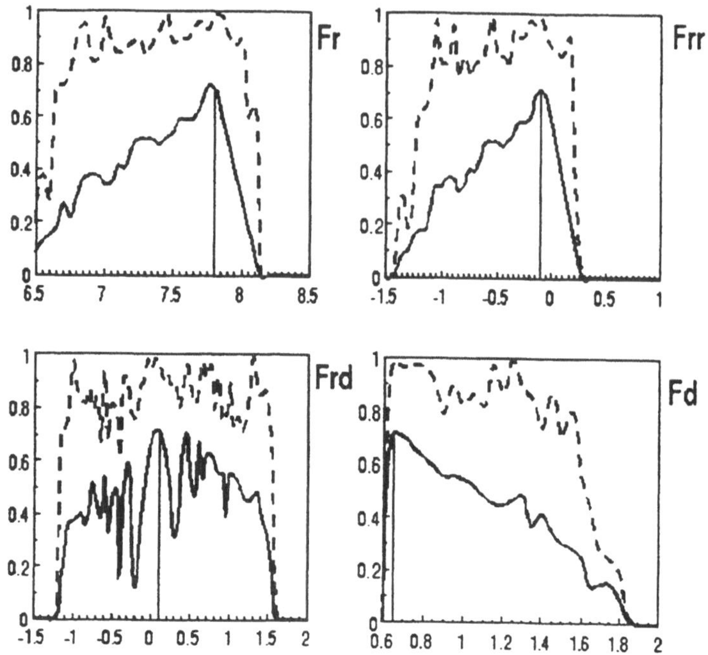

Figure 3: Projections to the force constant intervals

In distingush to the total fuzzy evaluation the isotope evaluation is illustrated as a stroked line. We can interprete these projections as evaluations on the assumed force constant intervals. A force field is called **fuzzy-optimal**, if the force constant combination has the greatest isotope evaluation under all combinations with a total fuzzy evaluation greater than 0.5. Finally, we get the following force field (in brackets the *LSE solution* is given):

$F_r^* = \mathbf{7.8}$ (*7.57*) mdyn/ Å, $F_{rr}^* = \mathbf{-0.1}$ (*-0.26*) mdyn/ Å,

$F_d^* = \mathbf{0.65}$ (*0.73*) mdyn Å $F_{rd}^* = \mathbf{0.05}$ (*-0.43*) mdyn.

To an analysis of the force constants the projections play an important role. Especially, we can give the following interpretation: If for $Fr = Fr^*$ the projection value is equal to α, than all force constant combinations (Fr^*, Frr, Frd, Fd) have at most the total fuzzy evaluation of α and for at least one combination the evaluation is equal to α. Analogous statements hold for the other force constants. However, we have to emphasize that these interpretations require the definition of cross product and projection with minimum and supremum operator, resp..

Now, we will investigate the total fuzzy evaluation near by the fuzzy optimal point (Fr^*, Frr^*, Frd^*, Fd^*). To do this, we consider the membership degrees of combinations (Fr, Frr^*, Frd^*, Fd^*), (Fr^*, Frr, Frd^*, Fd^*), (Fr^*, Frr^*, Frd, Fd^*) and (Fr^*, Frr^*, Frd^*, Fr) where Fr, Frr, Frd and Fd can take any value in the intervals I_r, I_{rr}, I_{rd} and I_d With other words, we consider the projections of the fuzzy relation B to Fr, Frr, Frd and Fd under the secondary condition that the values of the other force constants hold fixed. We will call these projections *View from the top*. In the case of the water molecule, the view from the top is illustrated in figure 4 by a continous line.

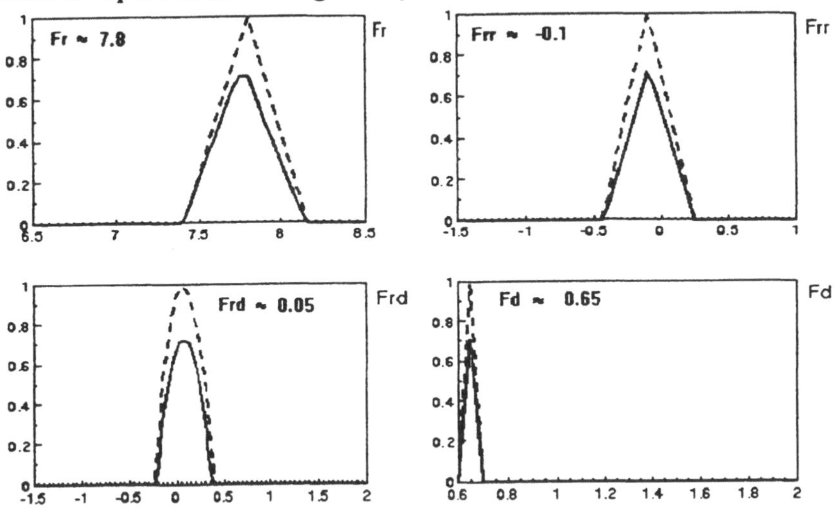

Figure 4: View from the top and fuzzy force constants for H_2O

Starting from the view from the top, we are able to specify the statement "*near by*" the optimal force field in an analytic way. If the projection curves are normalized, fuzzy numbers $\tilde{F}r, \tilde{F}rr, \tilde{F}rd$ and $\tilde{F}d$ on the force constant intervals I_r, I_{rr}, I_{rd} and I_d arise. From spectroscopical point of view, these fuzzy numbers can be interpreted as fuzzy force constants. Now, the consideration of fuzzy force constants into the NCA means to solve a fuzzy eigenvalue problem. The problem can be solved by α-cuts. Hence, fuzzy eigenfrequencies arise. The membership functions of the fuzzy eigenfrequencies are illustrated in figure 5 by continous curves. Comparing the fuzzy eigenfrequencies with the fuzzy observations of the experimental frequencies, marked in figure 5 by doted lines, we conclude that the fuzziness of the observations is reflected by the fuzziness of the force constants.

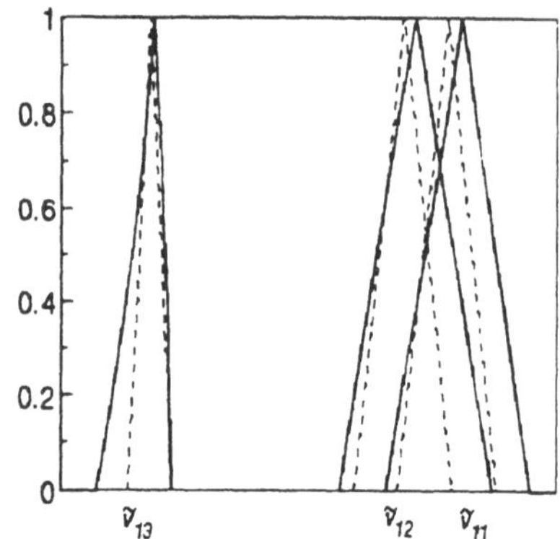

Figure 5: Fuzzy eigenfrequencies and fuzzy observations

Summarizing the results, the fuzzy approach to the NCA-model shows an interesting new way to the determination of physical usefull force fields. A main advantage of the approch is that theoretical as well as experimental uncertainty and expert knowledge about the band positions of the vibrations can be taken into consideration. A transfer of the approach to bigger molecules is of main interest to investigate the couplings between vibrations of different atomic groups. Especially, about the numerical values of coupling force constants vary vage informtion are available. The fuzzification of the coupling fore constants can be an interesting new way to get more informations about the force field of complex molecules like zeolites.

References

[McIntosh, Michaelian 1979] The Wilson GF Matrix Method of Vibrational Analysis, Part I- III,Canadian Journal of Spectroscopy 24, (1979)

[Jones 1976] Computer Programs for Infrared Spectrophotometry- Normal Coordinate Analysis. N.R.C.C. Bulletin No. 15, Canada (1976)

[Nibler, Pimentel 1968] Force constant Displays of Unsymmetric Molecular Isotopes of H_2O, H_2S, H_2Se, and HCCH
Journal of Molecular Spectroscopy 26, (1968)

[Kudra et al. 1993] A Fuzzy Model to the Refinement Problem of Water Force Constants, Preprint, (1993)

6.4. Fuzzy Elastic Matching of Medical Objects Using Fuzzy Geometric Representations

L. Köhler and P. Jensch

Abstract. The term fuzzy elastic matching describes an algorithm based on the combination of heuristic search and fuzzy heuristics. It interrelates two medical objects represented as wireframes or attributed skeletons. Parts of a medical object (i.e. coronary arteries) can be identified considering an abstract model. The fusion of different image modalities (i.e. liver angiographiy/ tomography) into a normalized one can be achieved using a common element. A new fuzzy algebraic and geometric framework enables us to introduce fuzzyness into object representations to treat their natural unsharpness.

1 Introduction

A prerequisite for the processing of medical objects is their representation as wireframe objects or attributed skeletons. Lateron these data structures can be visualised again as medical objects and can be given artificial attributes such as transparency of their surface to have an insight [Kö93] [MaKö93]. The Renderman language has defined a standard for modelling spatial objects and defining visualisation senaries. It's part of the operating system NeXTstep, so the management of wireframe objects is tightly coupled to other data modalities such as text, vector graphics, images, control elements or even sound, which leads to a uniform way of integrating them into multimodal applications [ME92]. The language Prolog guaranties both an easy way of representing and of searching in a large set of wireframe nodes. Parallel Prolog is able to do a heuristic search and our fuzzy arithmetic unit is able to implement the heuristic for directing the search. The compilation of prolog not only provides efficiency gains in processing but also essential information for efficiently representing data structures [Kö89].

2 Fuzzy Representation of Wireframes And Skeletons

We have to introduce fuzziness into classical wireframe representations to treat the unsharpness of natural objects.Fuzzy vectors, fuzzy edges and fuzzy angles substitute the crisp components of wireframes.

2.1 Semantics

The definitons [KruGebKla93] form the basis for formalizing the *semantics of fuzzy numbers, -vectors, -angles and -coordinate systems.*

Definition 1 Fuzzy Equivalence Relation. Let $\top$ be a t-norm. The mapping $E_\top : X \times X \rightarrow [0,1]$ is called an equivalence relation with regard to $\top$, if its reflexive, symmetric and transitive as follows

$$\top(E_\top(x,x'), E_\top(x',x'')) \leqq E_\top(x',x'')$$

The property *extensionality of fuzzy sets* to be defined now restricts fuzzy sets to be compatible to the underlying equivalence relation.

Definition 2 Extensionality. Let $E_\top$ be a fuzzy equivalence relation in X. A fuzzy set $\tilde{A}_X$ is called *extensional* if it has the following property:

$$\top(\mu_{\tilde{A}}(x), E_\top(x,y)) \leqq \mu_{\tilde{A}}(y).$$

In effect this interrelation has the essential implication that it's valid

- $E(x,x') = 1 \quad \Rightarrow \quad \mu_{\tilde{A}}(x) = \mu_{\tilde{A}}(x')$
- $E(x,x') \geqq \epsilon \quad \Rightarrow \quad \inf_{x:E(x,x')\geqq\epsilon} \{\mu_{\tilde{A}}(x)\} \geqq \top(\sup_{x:E(x,x')\geqq\epsilon} \{\mu_{\tilde{A}}(x)\}, \epsilon)$
- $\epsilon \leqq \epsilon' \quad \Rightarrow \quad \begin{cases} \inf_{E(x,x')\geqq\epsilon} \{\mu_{\tilde{A}}(x)\} \leqq \inf_{E(x,x')\geqq\epsilon'} \{\mu_{\tilde{A}}(x)\} \\ \sup_{E(x,x')\geqq\epsilon} \{\mu_{\tilde{A}}(x)\} \geqq \sup_{E(x,x')\geqq\epsilon'} \{\mu_{\tilde{A}}(x)\} \end{cases}$

By fixing one — for a normalized fuzzy set $\tilde{A}$ characteristic — μ-value the definition interval for all other μ-values is restricted to maintain the compatibility with the underlying equivalence relation:

$$\mu_{\tilde{A}}(x_1) = 1 \Rightarrow \inf_{E(x,x_1)\geqq\epsilon} \{\mu_{\tilde{A}}(x)\} \geqq \top(\sup_{E(x,x_1)\geqq\epsilon} \{\mu_{\tilde{A}}(x)\}, \epsilon) \geqq E(x,x_1)$$

The property *extensionality* now assures the fact, that every E-extensional fuzzy set subsumes a singleton of E so having at least its fuzzyness. We showed that an extensional mapping that obeys $E_{\top_1}(x,x') \wedge_\top \ldots \wedge_\top E_{\top'_n}(x,x') \leqq F_{\top'}(\phi(x),\phi(y))$ with $\top \leqq \top'$ maps extensional to extensional fuzzy sets so preserving the fuzzyness.

2.2 Fuzzy Number Calculation

For reasons of efficiency restricted classes of fuzzy numbers are used: LR-fuzzy numbers which are based on linear or exponential reference functions [DubPr80]: $L(z) = R(z) = \max(0, 1-z)$. The fuzzy extension is specified by the paramters α and β. LR-addition is defined as follows:

$$(a,\alpha,\beta)_{LR} + (a',\alpha',\beta')_{LR} = (a+a',\alpha+\alpha',\beta+\beta')_{LR}$$

To achieve notational consistency we introduce the following modification to LR-numbers. This has effect only to negative values.

Definition 3 Extended LR-Number Representation. A LR-number is represented by the tripel $A = (a,\alpha,\beta)_{LR}$ with the following semantics

$$\mu_{A=(a,\alpha,\beta)_{LR}}(x) = \begin{cases} L\left(\frac{a-x}{\alpha}\right) & \text{for } |a-\alpha| \leqq |x| \leqq |a| \\ R\left(\frac{x-a}{\beta}\right) & \text{for } |a| \leqq |x| \leqq |a+\beta| \end{cases}.$$

For negative LR-fuzzy numbers this leads to the modified representation $-\tilde{A} = (-a,-\alpha,-\beta)_{LR}$. The described fuzzy set remains the same.

μ-values the the 'additive'

Instead of defining LR-subtraction independently we derive its semantics from LR-addition and -negation. The treatment of positive and negative values is based on the 'pyramidal' evaluation of μ-values following the extionsion principle and determining the $\mu(z)$-value assignment to $z = x+y$ or $z = x-y$ *both* along the 'additive' diagonal. Conventional LR-subtraction can be modelled by subtracting a crisp and adding a zero fuzzy vector.

Theorem 4 Compensatory LR-subtraction. *Given the definitions of negative LR-numbers and LR-addition there is a semantics for LR-subtraction that delivers a solution to the equation*

$$\tilde{A} - \tilde{X} + \tilde{B} = \tilde{A} \quad \textit{with} \quad \tilde{X} = \tilde{B}$$

Compensatory LR-subtraction will be defined by

$$\begin{aligned}(a,\alpha,\beta)_{LR}-(a',\alpha',\beta')_{LR} &= (a,\alpha,\beta)_{LR}+(-a',-\alpha',-\beta')_{LR}\\ &= (a-a',\alpha-\alpha',\beta-\beta')_{LR}\end{aligned}$$

using the extended LR-notation for positive and negative LR-numbers.

Proof. The semantics of the difference fuzzy number $\tilde{A}-\tilde{B}$ is defined by the equation $\tilde{A}-\tilde{B}+\tilde{B}=\tilde{A}$ on the basis of fuzzy addition.

$$\begin{aligned}\mu_{(\tilde{A}-\tilde{B})+\tilde{B}}(x) &= \mu_{\tilde{A}}(x)\\ \mu_{(\tilde{A}-\tilde{B})+\tilde{B}}(x) &= \sup_{x=z+y}\min\{\mu_{\tilde{A}-\tilde{B}}(z),\mu_{\tilde{B}}(y)\}\\ \mu_{(\tilde{A}-\tilde{B})+\tilde{B}}(x_0) &= \mu_{\tilde{A}-\tilde{B}}(z_0)=\mu_{\tilde{B}}(y_0)=\mu_{\tilde{A}}(x_0)=\mu_{\tilde{A}+\tilde{B}}(x_0+y_0)\\ \mu_{\tilde{A}-\tilde{B}}(x_0-y_0) &= \mu_{\tilde{A}+\tilde{B}}(x_0+y_0)\end{aligned}$$

Evaluating the sup/min-operators on equivalent reference functions yields the equivalence of the function parameters $\mu_{\tilde{A}-\tilde{B}}(z)$ / $\mu_{\tilde{B}}(y_0)$ and of $\mu_{(\tilde{A}-\tilde{B})+\tilde{B}}(x)$ on certain z_0/y_0 values (along the diagonal $x=z+y$). Because of $\mu_{(\tilde{A}-\tilde{B})+\tilde{B}}(x)=\mu_{\tilde{A}}(x)$ this equality can be transfered to $\mu_{\tilde{A}-\tilde{B}}(x_0-y_0)=\mu_{\tilde{A}+\tilde{B}}(x_0+y_0)$. For addition the relation between x_0,y_0 and z yields:

$$\begin{aligned}\mu_{\tilde{U}+\tilde{V}}(z') &= \sup_{z'=x+y}\min\{\mu_{\tilde{U}}(x),\mu_{\tilde{V}}(y)\}\\ &= \sup_{z'=x+y}\min\{L_U\left(\tfrac{m_U-x}{\alpha}\right),L_V\left(\tfrac{m_V-y}{\beta}\right)\}\\ &= \sup_{z'=x+y}\min\{1-\tfrac{m_U-x}{\alpha},1-\tfrac{m_V-y}{\beta}\}\\ &= \mu_{\tilde{U}}(x_0)=\mu_{\tilde{V}}(y_0) \qquad ,x_0=m_U-\tfrac{\alpha}{\beta}(m_V-y_0)\\ &= \tfrac{m_U+m_V-z'}{\alpha+\beta} \qquad ,x_0=\tfrac{\beta m_U-\alpha m_V+\alpha z'}{\alpha+\beta}\end{aligned}$$

Calculating the fuzzy difference now leads to the definition of LR-subtraction. Substituting $x_0+y_0=x_0+x_0-z$ for z' and calculating the sum results in new dependencies for x_0 and $\mu_{\tilde{A}+\tilde{B}}(x_0+y_0)$ of $z=x_0-y_0$.

$$\begin{aligned}\mu_{\tilde{A}-\tilde{B}}(z=x_0-y_0) &= \mu_{\tilde{A}+\tilde{B}}(x_0+y_0)=\mu_{\tilde{A}+\tilde{B}}(x_0+x_0-z)\\ &= \tfrac{m_A+m_B-(2x_0-z)}{\alpha+\beta} \qquad ,x_0=\tfrac{\beta m_A-\alpha m_B+\alpha z}{\beta-\alpha}\\ &= \tfrac{(m_A-m_B)-z}{\alpha-\beta}\end{aligned}$$

This last formula uncovers the semantics of LR-subtraction:
$(a,\alpha,\beta)_{LR}-(a',\alpha',\beta')_{LR}=(a-a',\alpha-\alpha',\beta-\beta')_{LR}$ □

Theorem 5 Group Property. *The set of triagular LR-numbers together with addition establishes a Group.*
• $\tilde{A}+\tilde{0}=\tilde{A}$ • $\tilde{A}+(\tilde{-A})=\tilde{0}$ • $(\tilde{A}+\tilde{B})+\tilde{C}=\tilde{A}+(\tilde{B}+\tilde{C})$

Proof. For associativity/ existence of a neutral element see [DubPr80]. The existence of an inverse element has been shown above. □

2.3 Fuzzy–Vectors

The presented approach is a generalisation of the concept of [Ce87] and extends the applicability of triangular LR-fuzzy numbers to the multidimensional space. A fuzzy vector is defined componentwise by triangular fuzzy numbers. Its global membership function is defined by cummulating the membership function of each component using a Yager-t-norm.

Definition 6. Let $\tilde{A}_i, 1 \leq i \leq n$, be triangular fuzzy numbers of type LR, $\tilde{A}_i \hat{=} (a_i, \alpha_i, \beta_i), \tilde{A}_i \subseteq I\!R \times [0 \ldots 1]$. Let $X = (X_1, \ldots, X_n)^t$ be a vector with $X_i \in I\!R$. Then the *fuzzy-Vector* $\tilde{A} = (\tilde{A}_1, \ldots, \tilde{A}_n)^t$ is defined by its membership function $\mu_{\tilde{A}}(X) = \max\left(0, 1 - \sqrt{\sum_{i=1}^{n} \left(1 - \mu_{\tilde{A}_i}(X_i)\right)^2}\right)$

$$\mu_{\tilde{A}}(X) = \max\left(0,\ 1 - \sqrt{\sum_{i=1}^{n} \left(\frac{|a_i - X_i|}{spr_i}\right)^2}\right) \qquad spr_i = \begin{cases} \alpha_i; X_i \leq a_i \\ \beta_i; X_i \geq a_i \end{cases}$$

Rendering a 3-dimensional fuzzy vector results in an oblique ellipsoid. The mathematical properties of fuzzy numbers are valid for fuzzy vectors.

2.4 Fuzzyfying a Coordinate System

Restricting our attention to E-extensional fuzzy vectors we are able to introduce a fourth category of coordinate system transformations:

- scaling
- translation
- rotation
- fuzzyfying.

The effect of this extension will be seen considering fuzzy-angles.

Definition 7 Fuzzy Coordinate System. Let X^n be the space in which to consider fuzzy vectors and E be a fuzzy equivalence relation on X. Any E-extensional fuzzy vector in a fuzzy coordinate system will be given an additional index E in LR-notation: $\tilde{A}_E = (a, \alpha, \beta)_{LRE}$.

The *fuzzyfied coordinate system/ space* $\tilde{X}_E^n$ is obtained by subtracting from any of its fuzzy vectors

$$\tilde{X}^n = ((x_1, \xi_{1_L}, \xi_{1_R}), \ldots, (x_n, \xi_{n_L}, \xi_{n_R}))_{LR}$$

the zero singleton vector

$$\tilde{0}_S^n = ((0, \epsilon, \epsilon), \ldots, (0, \epsilon, \epsilon)_{LR} \quad \text{with} \quad \mu_{\tilde{0}_S}(x) = E(0, x)$$

so reducing its relative fuzzyness to

$$\tilde{X}_E^n = ((x_1, \xi_{1_L} - \epsilon, \xi_{1_R} - \epsilon), \ldots, (x_n, \xi_{n_L} - \epsilon, \xi_{n_R} - \epsilon))_{LRE}.$$

and increasing that of the coordinate system. Adding a zero singleton will decrease general fuzzyness.

A vector which represents a singleton will become a crisp vector indexed by E. Of any extensional fuzzy vector the common part of fuzzyness will diminish leaving just the rest and the index E. This will allow to treat fuzzy vectors as single quantities.

2.5 Fuzzy Angles

Fuzzy Angle Coordinates of a Fuzzy Vector

The representation of a fuzzy vector skeleton has to be invariant with respect to orientation in space. Therefore we will develop a special kind of a fuzzy polygon tree, based on a fuzzy angle definition.

Drawing lines through the origin and an arbitrary element of an α-cut we can determine its angle to the abscissa and the bounds of the interval of angles. Let $\Phi : \mathbb{R} \times \mathbb{R} \to [0, 2\,\pi]$ be a mapping that determines the angle between the abscissa and a vector and let $\sup \emptyset = 0$. Let $\tilde{A} = (\tilde{A}_1, \tilde{A}_2)$ be a twodimensional fuzzy vector with $\tilde{A}_i = (a_i, \alpha_i, \beta_i), \quad i \in \{1, 2\}$. Then the fuzzy angle $\tilde{\alpha}$ is a fuzzy-set in $[0, 2\,\pi]$, defined by

$$\begin{aligned} \tilde{\alpha} &= \Big\{ (\varphi, \mu_{\tilde{\alpha}}(\varphi)) \mid \varphi = \Phi(x, y), \quad (x, y) \in \mathbb{R} \times \mathbb{R} \Big\} \\ \mu_{\tilde{\alpha}}(\varphi) &= \sup_{(x,y) \in \Phi^{-1}(\varphi)} \min \{\mu_{\tilde{A}}(x, y)\} \end{aligned}$$

The mapping Φ is defined by

$$\Phi(x, y) = \arccos \left(\frac{x}{\sqrt{x^2 + y^2}} \right), \qquad y \geq 0$$

Any angle value is a member of a fuzzy angle if the line represented by it touches or cuts at least one α-cut of the fuzzy vector. The maximum α-value of these α-cuts determines the membership grade of the angle. Our representation of a fuzzy angle is a fuzzy number. The nonlinearity of the sine and cosine functions causes a fuzzy angle not to be a triangular fuzzy number. But if the spreads of the fuzzy vector are very small with respect to the distance to the origin, the fuzzy angle can be approximated by a triangular fuzzy number.

Let $\tilde{A} = (\tilde{A}_1, \tilde{A}_2)$, $\tilde{A}_i = (a_i, \alpha_i, \beta_i)_{LR}, i \in \{1,2\}$ be a fuzzy vector and let $\tilde{\alpha}$ the correspondence fuzzy angle with support $S(\tilde{\alpha}) = [\varphi_{min}, \varphi_{max}]$ and 'mean value' or apex angle φ. Now its possible to approximate $\tilde{\alpha}$ by

$$
\begin{aligned}
\tilde{\alpha} &\approx \left(\varphi, \varphi - \varphi_{min}, \varphi_{max} - \varphi\right)_{LR} \qquad \text{with} \\
\varphi &= \Phi(a_1, a_2), \\
\varphi_{max} &= \Phi(x_{s_{max}}, y_{s_{max}}) \\
\varphi_{min} &= \Phi(x_{s_{min}}, y_{s_{min}}) \\
\varphi \in [0, \tfrac{\pi}{2}[\quad &: \quad \left(x_{max}, y_{max}\right) = \left(a_1 - \beta_2 * \cos(\varphi), a_2 + \alpha_1 * \sin(\varphi)\right) \\
& \quad\quad \left(x_{min}, y_{min}\right) = \left(a_1 + \alpha_2 * \cos(\varphi), a_2 - \beta_1 * \sin(\varphi)\right)
\end{aligned}
$$

The coordinates $(x_{s_{max}}, y_{s_{max}})$ and $(x_{s_{min}}, y_{s_{min}})$ describe the points of intersection of the perpendicular to the line with angle φ through point (a_1, a_2) with the border of the α-cut for $\alpha = 0$. So the approximated fuzzy angle is a subset of the exact fuzzy angle .

Fuzzy Angles Relative to a Fuzzy Vector

Now we want to describe the orientation of a fuzzy vector with respect to another one. The idea is to determine a fuzzy angle of a fuzzy vector whose origin isn't a crisp point. We consider the set of all lines that cut the α-cuts of both fuzzy-vectors for a fixed α. If the spreads of the fuzzy vectors are non zero, then there exist exactly four lines that are tangential to the α-cuts. The extreme angles of these lines are the bounds to the α-cut of the fuzzy angle. The set of all α-cuts is a monotone decreasing series which defines the fuzzy angle completely.

Its obvious that this fuzzy angle definition is equivalent to the fuzzy angle of the difference vector with a crisp origin. The span of the fuzzy angle is determined by the sum of all spreads which also is the spread of the difference vector calculated by conventional subtraction. The real reason for taking into account the sum of all spreads is somewhat more complex.

A difference vector for determining the relative angle just results out of a coordinate system translation which locates the new origin in the apex of the reference fuzzy vector. Such a coordinate translation is achieved and has to be based on a crisp translation vector. If you transform by subtracting a fuzzy vector actually two transformations take place: the translation by the apex of the fuzzy vector and the extension of the coordinate system fuzzyness. Each crisp vector in the transformed coordinate system implicitly gets the fuzzyness of the transformation vector. The fuzzy angle definition is based on the difference of the fuzzy vector spreads — the spread of the difference vector — and reflects the relative angle of fuzzy vectors in a coordinate system of extended fuzzyness. But the angle of a crisp vector in a fuzzy coordinate system has a fixed fuzzyness determined by its own fuzzy spreads in normalized fuzzy space and that of the origin. Both equal the fuzzy extension of the transformation vector. The total fuzzyness of the relative fuzzy angle now is increased by twice these spreads which again amounts to the sum of all spreads.

The calculation of an angle between three fuzzy vectors is inconsistent using traditional fuzzy arithmetic but valid within the developed framework.

3 Fuzzy Processing of Artery Skeletons

3.1 Fuzzy Representation of Artery Model

We describe an orientation invariant representation of an artery model based on fuzzy vectors and -angles to identify arteries. A polygon $(p_1 \ldots p_n)$ with $p_i \in R^3$ will be defined invariant with respect to orientation. Each segment with endpoints p_i, p_{i+1} called primitive characterises it's own coordinate system. So the primitive p_{i+1}, p_{i+2} is defined as a triple of attributes (a, α, β), where a is the euclidian distance between the points p_{i+1} and p_{i+2}, α is the angle between the y-axis and the projection of (p_{i+1}, p_{i+2}) onto the xy-plane, and β is the angle between the x-axis and the projection of this primitive onto the yz-plane. This description was done in the local coordinate-system of the primitive (p_i, p_{i+1}). The next primitive (p_{i+2}, p_{i+3}) has to be defined in the local coordinate system of (p_{i+1}, p_{i+2}). Therefore we rotate the system (p_i, p_{i+1}) about the angles α and β.

The exact position of a vessel branchpoint is difficult to describe because of the diameters of the vessel segments. So we define such a branchpoint using a fuzzy vector centred inside the branch. It is determined by fuzzy mathematical morphology using a self extending, centering and moving circle shape. While extending the circle shape we evaluate the areas of replicated neighbour pixel partially corresponding to the round shape. This procedure simulates the vertical calculation of the partialvolumeeffect and also improves measurement of the diameter of an artery.

The spatial course of a blood vessel containing branch points can be approximated by a 3D-polygon using *fuzzy line segments* and *fuzzy angles*. These attributes are derived from fuzzy vectors. A fuzzy vector contains both the position of a segment endpoint — this is the apex of the vector centered inside the vessel — and the diameter of the vessel represented by its spreads. Now the representation of a complete vessel is the concatenation of vessel segments and branchpoints $bv = (ls_1, arc_1, \ldots ls_j, arc_j, \ldots)$. It describes the fuzzy skeleton of the artery as a string.

3.2 Fuzzy Skeleton Matching

Conventional string matching only deals with strings of discrete symbols. No numerical data or attributes are included. For pattern recognition it has been shown that injection of attributes into symbols makes it easier to handle noise or distortion and so increases recognition rates. Generally speaking, matching a finite string x of attributes with another one called y has to transform the attributes in x into those in y while constructing a minimum cost sequence.

Therefore fuzzy theory seems to be an adequate technique for determining similarities between unknown and enrolled patterns. Assuming $x \in X$ to be an unknown pattern out of a infinite set with it's fuzzy attributes, the recognition result is determined by the following equation using a similarity measure S: $j = \max_{j \in I} S_{ij}$ So we choose that pattern $y \in Y$ with the correspond possibility distributions $\pi_{i1}, \ldots, \pi_{in}$ that are most similar to that $\tilde{\mu}_{i1} \ldots \tilde{\mu}_{im}$ of pattern $x \in X$ with respect to the similarity-measure S_{ij}.

3.3 Similarity Measure for Fuzzy Vectors and Fuzzy Angles

The similarity between two fuzzy vectors A_1, A_2 should represent the distance between the apex $\vec{a}_1$ and $\vec{a}_2$ as well as the fuzziness represented by the spreads. In our implementation this similarity is able to measure both, the position as well as the diamter of a blood-vessel.

Definition 8 Fuzzy Similarity. Let V be the set of all fuzzy vectors. $s : V \times V \rightarrow [0, 1]$ is defined to be a fuzzy similarity if it's valid:

1. idempotence: $s(A, A) = 1 \ \forall A \in V$
2. commutativity: $s(A_1, A_2) = s(A_2, A_1), \forall A_1, A_2 \in V$
3. apex identity: $s(A_1, A_2) = maximum \Leftrightarrow \vec{a}_1 = \vec{a}_2$

4. spread monotony: $\forall A_1, A_2, A_3$
   - $\int_X(\mu_{A_1}(\vec{x}) \cap \mu_{A_2}(\vec{x}))d\vec{x} < \int_X(\mu_{A_1}(\vec{x}) \cap \mu_{A_3}(\vec{x}))d\vec{x} \Rightarrow s(A_1, A_2) < s(A_1, A_3)$ with $\vec{a}_2 = \vec{a}_3$, so the apex of A_2 and A_3 are the same.
   - $s(A_1, A_2) < s(A_1, A_3) \Rightarrow ||\vec{a}_1, \vec{a}_2|| \leq ||\vec{a}_1, \vec{a}_3|| \forall A_1, A_2, A_3 \in V$ with $\alpha_{A_{2,i}} = \alpha_{A_{3,i}}$ and $\beta_{A_{2,i}} = \beta_{A_{3,i}}$, so the spreads are the same.

Now consider: $sim(A_1, A_2) = \frac{\int_X(\mu_{A_1}(\vec{x}) \cap \mu_{A_2}(\vec{x}))d\vec{x}}{\int_X(\mu_{A_1}(\vec{x}) \cup \mu_{A_2}(\vec{x}))d\vec{x}}$

The similarity measure *sim* comes closest to meeting the requirements.

4 Conclusion

The developed fuzzy algebraic and geometric framework proved to be a valuable tool by which the natural unsharpness of medical objects can be treated. The interpretation of fuzzy values as heuristic values enriches traditional pattern matching approaches. Together with fuzzy geometric representations it enables to interrelate different but similar objects.

References

[Ce87] A. Celmiŋš, Least squares model fitting to fuzzy vector data. *Fuzzs Sets and Systems (FSS)* **22** (1987) 245–269

[Di90] P. Diamond, Metric spaces of fuzzy sets. *FSS* **35** (1990) 241–249

[DubPr80] D. Dubois, H. Prade: *Fuzzy Sets and Systems.* Acad. Press, 1980

[ME92] P. Jensch, L. Köhler, *Imaging by Multi-Modal Data Fusion and Fuzzy Inference.* Proc. VI Medit. Conf. Medical and Biol. Eng., pp 953-956, 1992

[Kö89] L. Köhler. *Implementierung eines OR-parallelen Berechnungsmodells für PROLOG.* Informatik Berichte, Bd 81, Universität Bonn, 1990

[Kö93] L. Köhler, T. Matschull, P. Jensch: *Fuzzy Elastic Matching von komplexen Drahtgittermodellen zur dreidimensionalen Darstellung der Leber.* In: Informatik Aktuell, Procs. 3. Dortmunder Fuzzy Tage, Springer, 1993

[KruGebKla93] R. Kruse, J. Gebhardt, F. Klawonn; *Fuzzy Systeme.*In: Leitfäden und Monograhien d. Informatik, Teubner, Stuttgart, 1993

[MaKö93] T. Matschull, L. Köhler, P. Jensch: *Räumliche Adaption von komplexen Drahtgittermodellen mittels Fuzzy Heuristiken am Beispiel der Leber.* Jahrestagung d. Ges. f. Biomedizinische Technik, Graz, 1993

6.5. FRED
Fuzzy Preference Decision Support System
A knowledge-based Approach for fuzzy multiattribute Preference Decision Making

Martina Wiemers

Abstract

Life insurance takes into account interests of all groups involved; on the one hand the interests of the insurance company and on the other hand the interests of the clients.

In this context the contract of insurance can be interpreted as a complex decision process which has to consider the different interests. In this article a method for checking an apply will be proposed. The method is motivated by fuzzifying an outranking approach. In the first step the apply will be prejudged by fuzzy inference and product rules and in a second step the analyzed alternatives of risk minimizing will be put in a hierachical order. The best alternative will be realized in a contract. The method is realized in the decision support system FRED (fuzzy preference decision support system) which improves a decision support system (DSS) with a fuzzy knowledge-based element to get a usable tool for checking life insurance applies. The system is developed on a personal computer.

1 The importance of checking applies in life assurance

The aims of an insurance company can be found in apply checking in a condensed form. Especially in apply checking many of high complex decision potentials can be found which can be interpreted as multidimensional decision spaces. Most of the estimations are of a qualitative nature. Each criterion of checking touches different aspects of the company. Therefore it is necessary to find an estimation of different criterions which can be divided into

estimations of the risk of the apply:

medical risk

moral risk

risk of speculation

estimations of the stock
stock expansion
stock selection
estimations of security
qualification of reinsurance.

Most of the estimations need personal experience and practice. The estimation founded on experience has to adjust with economic data like policy of the company, market conditions. The clerk in charge has to show himself to be flexible in order to find an estimation appropriate to the actual situation. From the point of view of the underwriter the aim of checking applies is to accept only contracts which agree with the aims of the company. Using a joint checking of applies the individual risks of an apply can be divided into desirable and undesirabel risks. Main element of the approach is the existence of the individual risk whose estimation is based upon experience. Consequently the approach cannot be done schematically but needs heuristic elements (sensitivity) which represents a characteristic element of checking applies. Furthermore a comprehensive knowledge about processes and use of processes is required for reproductable results especially with incomplete informations [Raestrup, O.(1992)]. Reading between the lines is often needed.

The decision about accepting or rejecting an apply has to execute in a short time. The decions requires superior use of qualitative estimations which are based upon an individual characteristic of each apply.

2 An integrated approach for checking applies

A method for representing personal and often intuitive elements for decison making is given by the theory of fuzzy sets. In fuzzy sets it is possible to use uncertain, fuzzy, and incomplete information. Advantages are given by using colloquial expressions in linguistic variables. In fuzzy sets the membership of an element to a set can be defined by a degree of membership. The situation of 'belongs to a set' which value 1 and 'belongs not to a set' with value 0 becomes effaced to a whole intervall [0, 1] [Jaeger, A.(1986)]. A fuzzy set over a discourse X is a set of pairs $F = \{ x, \mu_F(x) : x \in X \}$ with μ_F membership function of X. A linguistic variable is a quintuple (V, T, U, G, M) with V name of the linguistic variable, T term set of linguistic expression of V, G syntax rule for generating $x \in T$, M semantic rule. A linguistic variable can be illuminated by an example in the context of life assurance. The linguistic variable represents the medical risk of a person and includes terms like moderate handicap, medium handicap, strong handicap, intense handicap. Each term can be shown by trapezoid membership functions on the set of risk additions on the intervall from 0 to 150 %:

moderate handicap	(0, 15, 25, 30)
medium handicap	(20, 25, 50, 55)
strong handicap	(45, 50, 100, 110)
intense handicap	(90, 100, 150, 150)

The membership functions can be illumend in Fig. 1:

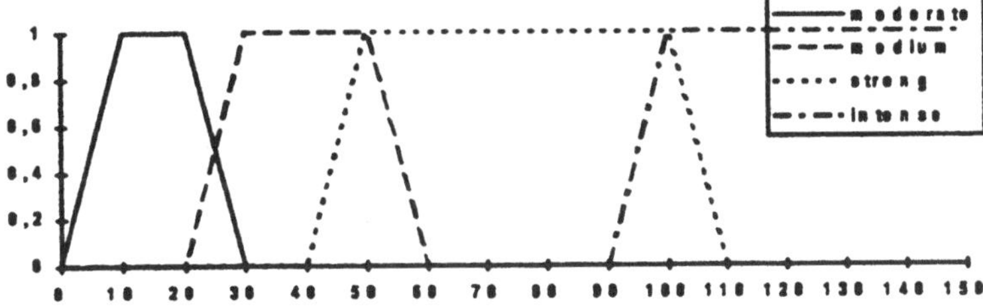

Fig.1.

Linguistic variables are used in fuzzy rules which are used for prejudgement of the apply. The fuzzy rules containt statistical (objective) and subjective knowledge. After the prejudgement the apply will be compared with other contracts to find out strong and weak points.
The decision support system FRED containt several steps:

registration of data and check of plausibility
prejudgement of the apply
comparison with contracts of reference
analysis of risk minimizing alternatives
recommendation

2.1 Registration of data and proof of plausibilities

In the first step data needed for the checking and contracting will be collected and checked. The data are given by the filled in form of the apply. Because of the regulations of law and clauses of the company the possibility of accepting the apply has to be decided about. Especially discrepancies with principles are checked.

2.2 Prejudgement

After first accept the apply will be estimated by different criterions. The criterions are the basis for the ranking in the next step. Because of rules representing statistical knowledge and personal experience with apply checking it will be easy to find an estimation. A fuzzy judgement will be possible by using approximate inference and production rules. Here linguistic variables can be used. An example for approximate reasoning will be illustrated by estimating overweight concering *age*, *weight* and *size* of a person. The linguistic variable *size* is formulated by the term set '*small, medium, large*', the variable *weight* by '*light, medium, heavy*' and *age* by '*young, medium, old*'. The parameters are given by:

Ling. Variable	Value	Parameter
weight (in kg)	light	40/0 50/1.0 70/1.0 80/0
	medium	60/0 70/1.0 100/1.0 110/0
	heavy	90/0 100/1.0 155/1.0 165/0
size (in cm)	small	140/1.0 150/1.0 164/1.0 174/0
	medium	156/0 165/1.0 180/1.0 190/0
	large	171/0 181/1.0 200/1.0 210/0
age (in years)	young	15/1.0 25/1.0 35/1.0 40/0
	medium	30/0 35/1.0 50/1.0 55/0
	old	45/0 50/1.0 55/1.0 60/1

The conclusion of the fuzzy rules gives hints about risk additions (*handicap*) of the individual risk. The linguistic variable *risk_addition* is given by:

Ling. Variable	Value	Parameter
handicap (in %)	moderate	0/0 15/1.0 25/1.0 30/0
	medium	20/0 25/1.0 50/1.0 55/0
	strong	45/0 50/1.0 100/1.0 110/0
	intense	90/0 100/1.0 150/1.0

Typical rules for finding the degree of additions with respect to overweight are:
If size small and weight light and age young, then handicap moderate (0.1).
If size small and weight medium and age young, then handicap moderate (1.0).
If size small and weight heavy and age young, then handicap moderate (0.5).

The membership degree of fullfilling the rule is given in brackets.
For example a medium handicap is given for a 28 years old , 160 cm tall man with a weight of 90 kg.

2.3 Comparison with Contracts of Reference

In the second step the apply will be compared with contracts of reference. Strong and weak points of the apply will be become plain. Mostly it will not be possible to accept or reject an apply with regard to all criterions. Concerning some criterions the apply will be accepted, concerning others the apply will be rejected. In this situation a compromise has to be found.
The proposed method is based upon an outranking approach which uses a degree of preference by a gradual (fuzzy) relation. The preference is modelled towards a claim. The degree of membership is integrated into the theory of possibility and necessity [Dubois, D., Prade, H.(1983), Zadeh, L.A.(1978)]. In the following the standard notation of fuzzy sets is used [Zadeh, L.A.(1973), Zimmermann. H.-J.(1991)].

Definition 1
Let F, G fuzzy numbers and P a fuzzy relation which gives the degree for $F \geq G$. The membership function of P is given by

$$\mu_P(F \geq G) = \sup_u \min\{\mu_F(u), \sup_{v \leq u} \mu_G(v)\} = \sup_{u,v,u \geq v} \min\{\mu_F(u), \mu_G(v)\}$$

μ_P is named possibility of $F \geq G$ [Zadeh, L.A.(1978), Baas, R.E., Kwakernaak, K.(1977), Dubois, D., Prade, H.(1983)].

Definition 2
Let F und G fuzzy numbers and N a fuzzy relation which gives the degree for $F > G$. The membership function of N is given by

$$\mu_N(F > G) = \inf_u \max\{1 - \mu_F(u), \inf_{v \leq u} \mu_G(v)\} = 1 - \sup_{u,v,u \leq v} \min\{\mu_F(u), \mu_G(v)\}$$

$$= 1 - \mu_P(G \geq F).$$

The relation N denotes the grade of necessity of strict dominance [Dubois, D., Prade, H.(1983)].
In the following suppose a multicriteria decision situation involving $X = \{x_1, ..., x_m\}$ set of alternatives and $K = \{k_1, ..., k_n\}$ set of criterions.

For each criterion k_j exists a linguistic variable Z_j with fuzzy numbers $\{Z_j^1, ..., Z_j^{l(j)}\}$ and $Z_j^1 < ... < Z_j^{l(j)}$.

For each criterion exists $opt_j = \{ max, min \}$. Each alternative is estimated concerning a criterion k_j by $z_{ij} \in Z_j$. For each alternative a vector $z^i = (z_{i1}, ..., z_{in})$ exists concerning all criterions $k_1, ..., k_n$ with $z_{i1} \in Z_1, ..., z_{in} \in Z_n$.

Example:

Suppose X a set of life assurance applies with X = {apply 1, apply 2, apply 3} and K criterions for checking with K = { medical risk, moral risk, risk of speculation, ability of reinsurance, estimation of stock value}.

	med. Risk	mor. Risk	Spec. Risk	Rein-surance	stock value
apply1	very high	medium	medium	high	very high
apply 2	medium	very low	low	high	high
apply 3	high	low	very low	low	medium
opt_i	min	min	min	max	max
Scale Z_i	very very low very low medium high very high very very high	very very low very low medium high very high very very high	very very low very low medium high very high very very high	very very low very low medium high very high very very high	very very low very low medium high very high very very high

The estimation of apply 1 is z^1 = (very high, medium, medium, high, very high).

2.3.1 Scales for each criterion

For modelling the linguistic variables four thresholds are given as indifference, preference, irritation and veto threshold (IS, PS, RS, VS). They are used for modelling the hypothesis '*towards criteria k_i is the value of the linguistic variable* Z_i^p *as good as* Z_i^q'. For expressing this hypothesis the possibililty relation can be used. Towards the claim 'Z_i^p *is as good as* Z_i^q' the change from a positive attitude to a negative attitude appears not abruptly but rather gradually when Z_i^p should become lager and larger than Z_i^q. The threshold for accepting or rejecting the claim is formulated by indifference and preference thresholds which can be interpreted as a fuzzy number. For each value $Z_i^q < Z_i^p + IS$ the membership degree towards the claim 'Z_i^p *is as good as* Z_i^q *with respect to criteria* k_i' is totally accepted with value 1 and for $Z_i^q > Z_i^p + PS$ the claim is totally unaccepted and gets the value 0. The membership degree for the preference claim decreases from 1 to 0. The membership function for the acception is supposed as $pref_i : Z_j \times Z_j \rightarrow [0,1]$ where pref is a fuzzy relation with $(opt_j = max)$:

$$pref_i(Z_i^p, Z_i^q) = \begin{cases} 1 & \mu_N(Z_i^q < Z_i^p + IS) = 1 \\ 0 & \mu_N(Z_i^q > Z_i^p + PS) = 1 \\ \text{decreasing} & \text{else} \end{cases}$$

Z_i^p constant.

Because of the exention principle is:

$$\mu_{Z_i^p + IS}(z) = \sup_{z = x+y} \min\{z_i^p(x), IS(y)\}$$

$$\Rightarrow \mu_N(z_i^q < z_i^p + IS) = \inf_u \max\{1 - \mu_{Z_i^p + IS}(u), \inf_{z \geq u} 1 - \mu_{Z_i^q + IS}(z)\} =$$

$$1 - \sup_{z \geq u} \min\{\mu_{Z_i^p + IS}(u), \mu_{Z_i^q + IS}(z)\} = 1 - \mu_P(z_i^q \geq z_i^p + IS).$$

Analogous there is a veto- and irritation (VS, RS) threshold. They are usable towards the claim 'Z_i^p *is worser than* Z_i^q *with respect to* k_j'. The membership function is supposed as $inko_i : Z_j \times Z_j \rightarrow [0,1]$ where (optj=max)

$$\text{inko}_i(Z_i^p, Z_i^q) = \begin{cases} 0 & \mu_N(Z_i^q < Z_i^p + RS) = 1 \\ 1 & \mu_N(Z_i^q > Z_i^p + VS) = 1 \\ \text{increasing} & \text{else} \end{cases}$$

Z_i^p constant.

The thresholds are independent of i and p and the increase of the degree is linear between the thresholds. A possiblity for a mathematical formulation of the increase is Zadeh's S-function [Giarrantano, J., Riley, G.(1989)] or Hellendoorn's generalization [Hellendoorn, H.(1990)].

2.3.2 Sets of alternatives

Now the scale values of the criterions are used for estimating the alternatives of each criterion. The quality of preference of the first step is used for constructing local relations of preference. The membership function is formulated towards the claim '*the alternative x_p is preferable to the alternative x_q with respect to criterion k_j*'. Based upon the preference function of the scales the preference situation of the alternatives will be defined. The local preference structure is a fuzzy relation defined as $P_j = (P_{pq}^j)$ with

$P_{pq}^j = \text{pref}(z_{pj}, z_{qj}), \forall p, q = 1, \ldots, m, j = 1, \ldots, n.$

Analogous the local incompensation relation is given by a membership function towards the claim '*alternative x_p is incompensatably worse than x_q with respect to criterion k_j*'. 'Incompensatably' means that a better value in one criterion cannot even out a worse value in another criterion for this alternative. The incompensation relation is defined as $I_j = (I_{pq}^j)$ with

$I_{pq}^j = (\text{inko}_j(z_{pj}, z_{qj}), \forall p, q = 1, \ldots, m, j = 1., \ldots, n.$

2.3.3. Constructing the concordance and discordance relation

The local relation will be aggregated to a global relation where all information about the alternatives with respect to each criterion is concentrated. The membership degree towards the claim '*alternative x_p is as good as alternative x_q with respect to all criterions*' can be represented as concordance degree C_{pq}. There all local values can be added to the fuzzy relation Cpq with $C_{pq} = \sum_{j=1}^{n} w_j P_{pq}^j$, p,q = 1,...,m, wj weight of criterion kj, $0 \le w_j \le 1, \forall j$.

The aggregation of the non-concordance called discordance leads to the membership degree towards the claim '*alternative x_p is incompensatibly worse than alternative x_q for each criterion k_j*'. The membership degree towards the claim '*alternative x_p is not compensatably worse then alternative x_q*' is more interesting. This claim is equal to '*alternative x_p is with respect to k_1 not incompensatably worse and with respect to k_2 not incompensatably and ... and with respect to k_n not incompensatably worse than x_q*'. By using the fuzzy NOT and AND operator the fuzzy incompensation relation (discordance relation) D_{pq} can be defined as:

$$D_{pq} = \prod_{j=1}^{n} (1 - I_{pq}^j)^{w_j}, \quad p,q = 1, ..., m, \text{ wj weight of criterion kj}, 0 \le wj \le 1, \forall j.$$

The preference relation can be formulated by combining the claim '*alternative x_p is as good as x_q for each criterion*' and '*alternative x_p is incompensatibly worse than alternative x_q for each criterion*' as $R = (R_{pq})$ with $R_{pq} = C_{pq} D_{pq}$. By using well known methods like Electre [Roy, B.(1991)] or Promethee [Brans, J.P.; Mareschal, B.; Vincke. Ph.(1985)] the preference relation can be used for decision recommendation.

3 Discussion

The presented method is a new way for apply checking in life insurance. The method allows to keep the identity and intrinsic properties of all criterions while at the same time taking into account the uncertainty of the evaluations. The concept of fuzzy sets is a useful method for modelling these uncertainties. Because of the well use of approximate inference in business administration problems this kind of inference is used for prejudgements of an apply. It suggests that much of the uncertainty which is intrisic in risk analysis is rooted in the fuzziness of the information and estimation. For instance preferred customers could be defined by degrees of membership concering each criterion of apply checking. This fuzzy estimations are taken up by the outranking approach for ranking the alternatives which can be evaluated for risk minimizing. Similar methods like Promethee are used in banking for checking credibility of customers. In insurance fuzzy concepts are not well known.Here imprecise statements have often transformed into 'all-or-nothing' rules. Fuzzy set theory can be used to provide a more flexible definition of insurableness and allows for some form compensation between the selected criteria.

References

Baas, R.E.;Kwakernaak, K. (1977): Rating and ranking of multiple-aspect alternatives using fuzzy sets, in: Automatica ,13, 1977, 47-58.

Brans, J.P.; Mareschal, B.; Vincke, Ph. (1985): A preference ranking organisation method. The Promethee method for MCDM, in: Management Science 31/6, 647-656.

Dubois, D.; Prade, H. (1991): Fuzzy sets in approximate reasoning, Part I: Inference with possibility distribution, Part II: Logical approaches, in: Fuzzy Sets and Systems, 40, 1991, 143-202, 203-244.

Dubois, D.; Prade, H. (1983): Ranking fuzzy numbers in the setting of possibility theory, in: Information Sciences, 30, 1983, 183-224.

Giarrantano, J.; Riley, G. (1989): Expert Systems: Principles and Programming, Boston, 1989.

Hellendorn, H. (1990): Closure properties of the compositional rule of inference, in: Fuzzy Sets and Systsms, 35, 1990, 163-183.

Jaeger, A. (1986): Bewältigung von Zielkonflikten mit Hilfe von graduellen Relationen, Schwellenwerten und Prävalenzen, Arbeitsbericht zur Ökonomathematik, No. 8602, Ruhr-Universität Bochum, 1986.

Raestrup, O. (1992): Prüfung des objektiven und subjektiven Risikos in der Lebensversicherung, in: Versicherungswirtschaft, 12, 1992, 146-749.

Roy, B. (1991): The outranking approach and the foundations of Electre methods, in: Theory and Decision, 31(1), 1991, 49-73.

Winkels, H.-M.; Wäscher, G. (1981): Konstruktion von Prävalenzrelationen, in: Fandel, G. et al. (eds.): Operations Research Proceedings 1980, Berlin, Heidelberg, 1981, 106-112.

Zadeh, L.A.(1978): Fuzzy sets as a basis for a theory of possibility, in: Fuzzy Sets and Systems, 1, 1978, 3-28.

Zadeh, L.A.(1973): Outline of a new approach to the analysis of complex systems and decision processes, in: IEEE, Vol. SMC-3, No.1, 1973, 28-44.

Zimmermann, H.-J. (1991): Fuzzy sets theory and its application, Boston, Dordrecht, Lancaster, 1991.

Authors

Bauer, P. Institut für Mathematik, Johannes Kepler Universität, A-4040 Linz, Austria

Bitterlich, N. Technische Universität Chemnitz-Zwickau, Fachbereich Elektrotechnik, Reichenhainer Straße 70, 09126 Chemnitz, Germany

Bocklisch, S.F. Technische Universität Chemnitz-Zwickau, Fachbereich Elektrotechnik, Reichenhainer Straße 70, 09126 Chemnitz, Germany

Brahim, K. Universität Stuttgart, IPVR, Breitwiesenstraße 20-22, 70565 Stuttgart, Germany

Dierkes, S. Universität Dortmund, Lehrstuhl Informatik I, Otto-Hahn-Straße 16, 44221 Dortmund, Germany

Felix, R. Fuzzy Demonstrations-Zentrum Dortmund, Joseph-von-Fraunhofer-Straße 20, 44227 Dortmund, Germany

Freksa, C. Universität Hamburg, FB Informatik, Vogt-Kölln-Straße 30, 22527 Hamburg, Germany

Gebhardt, J. Technische Universität Braunschweig, Institut für Betriebssysteme und Rechnerverbund, Bültenweg 74/75, 38106 Braunschweig, Germany

Glesner, M. Technische Hochschule Darmstadt, Institut für Datentechnik, Karlstraße 15, 64283 Darmstadt, Germany

Gottwald, S. Universität Leipzig, Institut für Logik / Wissenschaftstheorie, Augustusplatz 9, 04109 Leipzig, Germany

Halgamuge, S.K. Technische Hochschule Darmstadt, Institut für Datentechnik, Karlstraße 15, 64283 Darmstadt, Germany

Heinsohn, J. Deutsches Forschungszentrum für Künstliche Intelligenz GmbH, Stuhlsatzenhausweg 3, 66123 Saarbrücken, Germany

Hellendoorn, H. Siemens AG, ZFE ST SN 4, Otto-Hahn-Ring 6, 81739 München, Germany

Höhle, U. Bergische Universität Gesamthochschule Wuppertal, Fachbereich Mathematik, Gaußstraße 20, 42097 Wuppertal, Germany

Hopf, J. Max-Planck-Institut für Informatik, Im Stadtwald, 66123 Saarbrücken, Germany

Jensch, P. Universität Oldenburg, Institut für Angewandte Informatik, Postfach 2503, 26111 Oldenburg, Germany

Klawonn, F. Technische Universität Braunschweig, Institut für Betriebssysteme und Rechnerverbund, Bültenweg 74/75, 38106 Braunschweig, Germany

Klement, E.-P. Institut für Mathematik, Johannes Kepler Universität, A-4040 Linz, Austria

Knappe, H. Schröder airporttechnik, Mittenheimer Straße 74, 85764 Oberschleißheim, Germany

Köhler, L. Universität Oldenburg, Institut für Angewandte Informatik, Postfach 2503, 26111 Oldenburg, Germany

Kretzberg, T. Fuzzy Demonstrations-Zentrum Dortmund, Joseph-von-Fraunhofer-Straße 20, 44227 Dortmund, Germany

Kudra, M. Universität Leipzig, Institut für Physikalische und Theoritische Chemie, Linnéstraße 3, 04103 Leipzig, Germany

Leikermoser, A. ARS Ges.m.b.H., A-5020 Salzburg, Austria

Mari, A. Technische Hochschule Darmstadt, Institut für Datentechnik, Karlstraße 15, 64283 Darmstadt, Germany

Meyer-Gramann, K.-D. Bodan IMS GmbH, Postfach 1309, 88671 Markdorf, Germany

Moser, B. Institut für Mathematik, Johannes Kepler Universität, A–4040 Linz, Austria

Nauck, D. Technische Universität Braunschweig, Institut für Betriebssysteme und Rechnerverbund, Bültenweg 74/75, 38106 Braunschweig, Germany

Palm, R. Siemens AG, ZFE ST SN 43, Otto–Hahn–Ring 6, 81739 München, Germany

Reusch, B. Universität Dortmund, Lehrstuhl Informatik I, Otto–Hahn–Straße 16, 44221 Dortmund, Germany

Runkler, T.A. Technische Hochschule Darmstadt, Institut für Datentechnik, Karlstraße 15, 64283 Darmstadt, Germany

Sander, W. Technische Universität Braunschweig, Institut für Analysis, Pockelsstraße 14, 38106 Braunschweig, Germany

Temme, K.-H. Universität Dortmund, Lehrstuhl Informatik I, Otto–Hahn–Straße 16, 44221 Dortmund, Germany

Wehner, M. Fuzzy Demonstrations–Zentrum Dortmund, Joseph–von–Fraunhofer–Straße 20, 44227 Dortmund, Germany

Wiemers, M. Leineweberstraße 41, 45486 Mülheim a.d. Ruhr, Germany

Zell, A. Universität Stuttgart, IPVR, Breitwiesenstraße 20–22, 70565 Stuttgart, Germany

Zimmermann, H.-J. Institut für Wirtschaftswissenschaften, RWTH Aachen, Templergraben 64, 52062 Aachen, Germany

Index